An Optimization Primer

Springer
New York
Berlin
Heidelberg
Hong Kong
London
Milan
Paris
Tokyo

John Lawrence Nazareth

An Optimization Primer

On Models, Algorithms, and Duality

With 28 Illustrations

 Springer

John Lawrence Nazareth
Professor Emeritus
Department of Mathematics
Washington State University
Pullman, WA 99164-3113
USA
and
Affiliate Professor
Department of Applied Mathematics
University of Washington
Seattle, WA 98195
USA
nazareth@amath.washington.edu
http://www.math.wsu.edu/faculty/nazareth

Mathematics Subject Classification (2000): 49-xx, 90-xx

ISBN 0-387-21155-1   Printed on acid-free paper.

Printed in the United States of America.   (EB)

9 8 7 6 5 4 3 2 1   SPIN 10968901

Springer-Verlag is a part of *Springer Science+Business Media*

springeronline.com

To Abbey

PREFACE

Optimization is the art, science, and mathematics of finding the "best" member of a finite or infinite set of possible choices, based on some objective measure of the merit of each choice in the set. Three key facets of the subject are:

- the construction of optimization *models* that capture the range of available choices within a feasible set (constraint structure) and the measure of merit of any particular choice in a feasible set relative to its competitors (objective function value);

- the invention and implementation of efficient *algorithms* for solving optimization models (finding optimal solutions);

- a mathematical principle of *duality* that relates optimization models (and algorithms) to one another in a fundamental way. Duality cuts across the entire field of optimization and is useful, in particular, for identifying *optimality conditions*; i.e., criteria that a given member of a feasible set must satisfy in order to be an optimal solution.

This booklet provides a very gentle introduction to the above topics. We try to achieve a balance in our presentation between the *specific details of a representative set* of optimization models, algorithms, and their corresponding duality, on the one hand, and the *big picture* of optimization, as summarized by visual schematics and associated commentary, on the other. Our hope is that the reader will thereby obtain a good intuitive sense of the area encompassed by optimization—its essential character and broad scope, as well as a historical perspective—and be motivated to embark on a much more comprehensive exploration of the subject. The broad themes of our introductory booklet will continue to resonate, and they will assume a greater significance, during the course of further, in-depth study of optimization.

Our optimization primer is written for two main classes of readers.

Firstly, it is intended for *college students* taking an introductory course or a sequence of courses in optimization that cover the areas traditionally designated as network-flow programming, linear programming, and nonlinear programming. The purpose of our primer is to provide a broad and unified perspective on these areas, in a self-contained format. Thus it should be used as a *supplement* or *front end* to one of the many currently available introductory textbooks on optimization that treat the subject more comprehensively. If an introductory course begins with some refresher linear algebra and/or calculus, then what is often a tedious initiation can be considerably enlivened by accompanying it with material selected from this booklet. Alternatively, the booklet can be assigned as independent reading to be completed during the first few weeks of instruction, in order to provide background and motivation. Our primer will also be a useful source of motivation for *high-school students* embarking on their studies in mathematics and science, and thus it will be of interest to *high-school teachers* in these areas.

Secondly, this booklet is intended for the *general reader* who wants to obtain an overall sense of the field of optimization, but may possess only very rudimentary (or perhaps no) knowledge of linear algebra and calculus. To make the presentation as accessible as possible, sections that require a little more mathematical background are in a smaller print. Such material can be quickly skimmed (or entirely omitted) without disrupting the main flow of the discussion. Thus all that is required in order to understand a substantial portion of the booklet is that the reader be able to follow a logical argument, be unafraid of computers, and be not uncomfortable with the occasional use of an algebraic symbol in place of a number. Among general readers that are mathematically more sophisticated, this booklet will be of particular interest to *scientists, engineers, economists, planners, and business managers* who want to satisfy their curiosity about the increasingly popular subject of optimization and put its techniques to practical use within their areas of endeavour.

In addition, this introductory booklet will be of value to *instructors at colleges and universities* who are already specialists in optimization, because it offers a blueprint for restructuring a traditional optimization curriculum. In teaching optimization, primacy should be restored to network flows, a very accessible subject that can (and should!) be approached from both *combinatorial* and *continuous* perspectives. The combinatorial perspective permits many delightful optimization models, algorithms, and their associated duality theory to be introduced immediately and simply, with almost no prerequisite mathematical or computational background. The continuous perspective, which derives from the fact that many network models have associated equivalent linear programs, provides a natural avenue to the wider subject of linear programming (LP), its underlying duality theory, and its main solution engine—Dantzig's simplex method. The adaptation of the latter to network structures yields an especially elegant and powerful algorithm. The subject of linear programming must then be revisited, in light of the 1984 Karmarkar algorithmic revolution, in order to introduce more recent interior-point LP ideas that are rooted in techniques of nonlinear programming. Finally, the enveloping field of nonlinear programming, in particular, differentiable nonlinear optimization and equation solving, can be presented in greater generality. Our booklet follows this natural progression of development of optimization leading from network flows, through linear programming (including its networks specialization), to nonlinear programming and beyond. An optimization curriculum organized along these lines could use our primer *as the main text for an initiating, self-contained minicourse.* It could proceed at a leisurely pace, meeting once a week, for example, over a ten-week quarter, and covering about one chapter per week; or, more intensively, meeting twice or thrice a week, and covering two or three chapters each week, thereby completing the material in a month. This course would provide motivation for the subject and the entry point to other in-depth courses on optimization, preferably sequenced as outlined above.

The organization of our primer is as follows:

In Chapter 1, we begin with some very simple motivating examples of optimization problems in situations that are close to home.

Next, in Chapter 2, we introduce a quintessential optimization problem: finding the maximal flow in a network. We call this problem "quintessential" because it provides a truly exceptional window on three key aspects of the optimization field—models, algorithms, and duality—that permits each to be introduced in a simple and beautiful way, and with a minimum of required background.

In Chapter 3, we elaborate on duality in the setting of two other key network flow problems—matching and covering—that builds in a very natural way on the maximal flow development of Chapter 2.

Chapters 2 and 3 provide a springboard for a broad overview of network flows in the form of a schematic tree and associated commentary, which is the topic of Chapter 4.

Chapter 5 begins our discussion of linear programming and its associated duality theory.

Chapter 6 then describes the golden age of modeling-and-algorithmic development in optimization, which was initiated by G.B. Dantzig over half a century ago. In particular, it introduces the main solution engine of the subject—the simplex algorithm—along with its elegant specialization to the case of network flows.

Chapter 7 provides a window on the 1984 interior-point algorithmic revolution that was set in motion by a seminal contribution of N. Karmarkar. We describe two key ideas: affine scaling and the central path of a linear program. Many interior-point algorithms, implicitly or explicitly, combine affine scaling and centering techniques, and they are outlined in the concluding section of this chapter.

Chapter 8 describes the embracing subject of nonlinear programming, in particular, differentiable optimization and equation solving, and provides a schematic portrait of its main problem areas.

In Chapter 9, we consider resource planning problems that exhibit a mix of network and linear programming characteristics across a time axis. (Optimization problems that arise in practice often have such *mixed characteristics* and are inherently dynamic.) We begin our discussion with a simple timber harvesting problem, and continue with a more elaborate example involving rangeland planning. Optimization models of this type, which are identified by the acronym D_LP, are intuitively accessible and very broadly applicable to natural and renewable resource decision-making situations. We also describe PC software that can be used as an educational tool for obtaining hands-on experience with the formulation and solution of D_LP resource planning models.

Finally, in Chapter 10, the rangeland planning model of the preceding chapter provides a platform for presenting the "big picture" of optimization in the form of a third visual schematic, a "cubist portrait" of the field that brings our booklet to a conclusion.

As stated earlier, we have tried to strike a balance between *specific details* of optimization models, algorithms, and duality (Chapters 1–3, 5–7 and 9) and the *big picture* of optimization as summarized by visual schematics and associated commentary (Chapters 4, 8, and 10). Our primer should be used as a front end for more standard offerings in network-flow programming, linear programming, and nonlinear programming, which will then provide greater comprehensiveness and clarity. The motivated reader who has acquired the necessary background knowledge in optimization in this way and wishes then to embark on more advanced study will find much more detail on topics selected for this primer in our two recent Springer-Verlag monographs, Nazareth [2003], [2001]. The former provides a comprehensive discussion of the optimization models and algorithms introduced in Chapters 5 through 8. The latter describes the modeling/algorithmic approach introduced in Chapters 9 and 10 and a variety of application areas (and it also contains an executable demonstration computer program on CD-ROM). Many open avenues for research, development, and application are highlighted in these two monographs.

Acknowledgments: Without singling out any individuals in particular and thus avoiding sins of omission, I conclude by thanking, in general, my teachers, colleagues, and friends within the optimization and numerical analysis communities, from whom I've learned much, and with whom I've had the pleasure of interacting over the years.

I gratefully acknowledge the Department of (Pure and Applied) Mathematics at Washington State University, Pullman, for providing a balanced research and teaching environment and, more importantly, a flexible academic appointment that made this and other writings possible.

And I thank my affiliate department of Applied Mathematics at the University of Washington, Seattle, for generously lending a helping hand.

JLN, Bainbridge Island (January 2004)

TABLE OF CONTENTS

COMMENTS ON STYLE AND PRINT SIZE

In seeking a guiding light for expository writings such as this primer, one can do no better than quote from the preface of Vašek Chvátal's classic on linear programming (Chvátal [1983]):

> For me, writing this book was a long and enjoyable exercise in exposition. I would return again and again to Paul Halmos' essay on writing mathematics whenever I needed a bit of extra stimulation to carry on. Speak to someone, says Halmos; and I tried to follow his advice, by always keeping a few of my students in mind. At the same time, however, I could not help thinking of my friends and colleagues, excellent mathematicians who never got around to learning linear programming. For them, I tried to make the text as appetizing as I could by not making simple ideas sound esoteric. Furthermore, the exposition is accessible to readers with a minimal mathematical background: a reasonably bright and motivated high school student could follow and understand.

Our booklet on optimization has been written in the spirit of the foregoing remarks.

We have also borrowed Chvátal's blueprint for addressing the needs of readers with varying backgrounds. Sections that require a little more (elementary) mathematical background use a smaller print—they are literally harder to read—and they can be quickly skimmed, or even entirely omitted by the general reader, without losing the main thread of the discussion. Again quoting from the preface of Chvátal [1983]:

> Note that throughout the book, passages in small print may be skipped on the first reading without loss of continuity. However, small print does not signify inferiority; quite the contrary: a sophisticated reader may enjoy these passages more than the rest of the text.

1
Simple Motivating Examples

Optimization is the art, science, and mathematics of finding the "best" or *optimal member* of a finite or infinite set of possible choices (called the *feasible set*), where the selection is based on some *objective function value* or measure of merit of each choice in the feasible set. Here are four exceedingly simple, but nevertheless instructive examples of optimization problems found in situations close to home. Read as many (or as few) as you need for purposes of initial motivation.

1.1 Shopping for Food

Suppose you wish to buy a pound of the cheapest seafood available at your local supermarket on any particular day. The "set of possible choices," or feasible set, comprises the dozen or so items on display at the seafood counter, and the price tag for each choice, i.e., the posted cost per pound, is its "measure of merit," or objective function value. Making the best or optimal choice is trivial—just pick the seafood item with the smallest price tag! Yet, even in this simple setting, a decision-making problem can quickly become more involved. Suppose you are also concerned with the nutritive value of the fish, say its content of protein per pound, this information being posted alongside the price tag. Suppose that you must satisfy a minimum daily requirement (MDR) on seafood protein, and again you wish to make your purchase as cheaply as possible. Now you must decide not only on the type of seafood but also the amount to be purchased. A little thought might lead you to choose the particular seafood item for which the *ratio of cost per pound to protein per pound* is as small as possible, and then to buy as much of that item as you need to satisfy your constraint (MDR) on protein.

In order to demonstrate that this intuitive solution does indeed yield the optimal choice, it is helpful to formalize the optimization problem in some suitable way; and such formalization becomes a necessity as the decision task becomes more complicated, for example, when minimum daily requirements on several different nutrients are simultaneously imposed. (An optimal choice in this case may require the simultaneous purchase of suitable amounts of *more than one* type of seafood.) We see that the need has arisen, almost immediately, to *model* our decison-making problem in some appropriate manner, and then to devise a suitable solution technique, or *algorithm*, to extract the optimal solution from the model. More on this example in Chapter 5.

1.2   Watering the Garden

A small lake at the boundary of your backyard acreage is a source of water for a vegetable and fruit garden. The water flows via two pipe junctions J1 and J2, to the demand point, say D, where water is needed for the garden. A large-bore pipe, which is capable of carrying up to 50 gallons of water per minute at normal pressure, connects the source point at the lake, say S, to the junction J1, and a similar pipe connects the junction J2 to the demand point D. Small-bore pipes, capable of carrying up to 10 gallons per minute, connect S to J2, J1 to D, and J1 to J2. All pipes have one-way shutoff valves, so the flow can travel in only one direction. We can model this setup by the network shown in Figure 1.1, where the arrows indicate the direction along which water can flow and the numbers indicate the maximum amount of flow per minute that is possible on each pipe at normal pressure.

Note that the largest flow that can be sent out of S is 60 gallons per minute, and a similar amount can be received at D. However, the constraints that the flow of water must balance at J1 and J2—the amount of water going into each pipe junction must equal the amount going out—and the smaller capacities of some of the pipes at these junctions make it impossible to send 60 gallons per minute from S to D. On the other hand, 10 gallons per minute can obviously be sent from S to D through the network, for example, on the pipes S to J1 and J1 to D, with no flows needed on other pipes. *What is the maximum amount of flow of water (per minute) that can be sent from the lake to the garden?* A brief examination of the model network in Figure 1.1 will reveal that the answer is 30 gallons per minute, which is achieved by sending 10 gallons per minute from S to J2 and 20 gallons per minute from S to J1, with the constraints on the balance of flow of water at J1 and J2 determining the associated flows on all

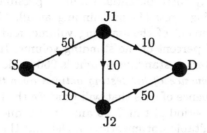

Circles are called nodes, and the connections are called links or arcs. Arrows represent the directions of flow.

FIGURE 1.1. Water distribution network

other pipes—10 from J1 to D, 10 from J1 to J2, and 20 from J2 to D. Any amount of total water flow from S to D between 0 and 30 gallons per minute is feasible on the network, i.e., there is an infinite number of feasible choices. The measure of merit or objective function value associated with each choice of flow through the network is just the total amount of the flow out of S (or, equivalently, into D). In this example, it is easy to find the optimal flow by inspection. But when this simple example is scaled up to a more realistic network model with many more nodes and arcs, the task of finding an optimal solution is impossible to perform by inspection and requires a suitable optimization algorithm. We will return to this example in the next chapter.

1.3 Chopping Wood

Next door to your home is a family farm, where timber has been grown and harvested for many decades on a portion of the land. Traditionally, this was done on a 30-year rotation, i.e, the timber was totally harvested 30 years after planting, and then the land was replanted. The new generation of the family, now in charge of the day-to-day operation of the farm, wants to evaluate the possibility of thinning before the end of the rotation period. The benefits of thinning are that small or diseased trees, which would normally die and have their timber lost before the rotation age, can be harvested. In addition, after a thinning, more light and nutrients are available for the remaining trees so they can grow to a larger diameter and thus achieve a greater log value. Thinning is to be permitted when the stands are 10 years old and again at 20 years. To simplify the instructions given to the harvesting crew, only three potential thinning possibilities are considered at the two harvesting ages: (1) no thinning at all, (2) light thinning to remove about 20 percent of the standing volume, and (3) heavy thinning to remove about 35 percent of the standing volume. Initially, the stand is of age 10 and the average standing volume is 260 cubic meters per hectare. *What is the best sequence of harvesting actions over the 30-year rotation,* i.e., what is the sequence of cuts that will produce the largest total amount of timber over this period of time? To answer this question again requires the building of a suitable optimization model and the invention of an algorithm (or the application of an existing algorithm) to find the optimal solution of the model. More on this example in Chapter 9; in particular, see the first note of Section 9.4 at the end of that chapter.

1.4   Going Fishing

Finally, let us return to the lake at the boundary of your property and go fishing in a small boat. The water of the lake is murky, and little can be seen below its surface. *Suppose you want to find the deepest point of the lake*, where the largest fish are known to lurk. You have an anchor on a long rope that can be used to measure depth at any location, but each depth sample requires effort that is not inconsiderable; i.e., it has an associated "informational cost." One approach to finding the deepest point is to place an imaginary grid over the surface and measure (sample) the depth at each point of the grid, as defined by its E-W and N-S coordinates. With a fine enough grid, one would obtain a good estimate of the location of the deepest point, but this approach requires many depth samples and thus a great expenditure of effort; i.e., it is very costly in terms of its informational needs. We must do better!

Let us *idealize* this situation and assume that the bottom of the lake is undulating but smooth, the water surface is perfectly flat, and the boat can remain stationary at any chosen point of the lake, where the anchor can be dropped in order to provide the depth at that point. (By moving the boat by small increments from the current location, in the E-W and N-S directions, and taking new depth samples, one can also obtain information about the slope of the bottom, formally called the gradient at the current location.) Your task is to use the information that *you choose to gather* at the current location of the boat, along with the information that you may have obtained at any previous locations, in order to move the boat to a new surface location, a better estimate of the deepest point. Then the process can be repeated, or iterated. Its aim is to converge to the point of the lake's surface where the water is deepest, and at low total informational cost, i.e., *by taking as few depth/slope samples as possible.* In the idealization, or optimization model, as defined by the foregoing assumptions, the feasible set of choices consists of the infinite number of locations on the bounded surface of the lake beneath which the optimal solution (deepest point) on the bottom could lie; and the objective function value, or measure of merit, of each point is the depth of the lake at that point. One seeks an algorithm for moving the boat to a sequence of locations (successive iterates) that converge to the optimal solution, and one wants the algorithm to be as efficient as possible in terms of its informational requirements (total number of depth/slope samples). More on this example in Chapter 8.

1.5   Summary

Without straying far from home, we have found a variety of simple examples of optimization problems, and these examples can easily be "scaled up" to a more realistic size. It then takes little further imagination to appreciate the fact that *optimization problems*, ranging from simple to very complex, *abound in every field of human endeavour.* They are especially prevalent, dressed in more mathematical clothing, within the many disciplines of the sciences and engineering.

We will return to the simple examples of this chapter and also encounter other more realistic examples of optimization problems in subsequent chapters of our primer.

2

A Quintessential Optimization Problem

Networks for distributing products or services are ubiquitous in modern life, and our everyday intution is already very well developed in this context. Thus networks provide an excellent vehicle for introducing optimization. Examples of network application areas are:

- transportation and/or distribution networks;

- communication networks;

- hydraulic networks;

- networks defined by the states of a discrete system that varies over time; see Chapter 9.

Finding the largest flow of products or services that can be sent through a network, from a source to a destination, is a *quintessential* problem of optimization, because it enables three key facets of the subject that are addressed in this booklet—models, algorithms, and duality—to be introduced simply and directly, and with a bare minimum of mathematical and computational background. We will consider each facet in turn.

2.1 Models

Consider the network depicted in the maximal water-flow example of Section 1.2 and its associated terminology of nodes and arcs; see Figure 1.1.

It is evident that the same network could be used instead to describe a simple instance of a maximum flow problem for any one of the network application areas itemized above.

Thus, in a *transportation and/or distribution network*, the nodes could represent cities, warehouses, ports, etc., and the arcs could represent road links, railroads, shipping routes, airline routes, etc. A variety of "products" could be transported on arcs, and the capacity of each arc would be the maximum amount of product that can be transported on it. For instance, S, J1, J2, and D in the network depicted in Figure 1.1 could represent four cities linked by airline routes corresponding to the arcs of the network. The stated capacities on the arcs, namely, 50 on the two arcs S to J1 and J2 to D, and 10 on the other arcs, could represent the *number of airline seats* that are available for daily booking on the corresponding routes. (Instead of units of flow being gallons per minute, the units are now passenger seats per day.) *What is the maximum number of passengers that can be booked to fly from city S to city D, assuming that all connections can be made at cities J1 and J2?* Clearly the answer is, again, 30 passengers per day, which is the solution of the previous maximum-flow of water problem associated with the network.

Similarly, in a *communication network*, the nodes could represent telephone exchanges, satellites, Internet servers, etc., and the arcs could represent cables, microwave relay links, etc. Again, the network of Figure 1.1 could be used to define a simple maximum flow problem in this area. Analogous observations apply to the other network application areas.

The modeling aspect of optimization within this context is thus immediately apparent, and *at any scale*, ranging from toy problems to large-scale, practical problems. The latter can be obtained simply by scaling up, i.e., by adding more nodes and arcs. A wide variety of practical application areas are thus immediately on hand within the setting of the network areas itemized in the list at the start of this chapter. Thus, without more ado, let us move on to a consideration of algorithms for solving them.

2.2   Algorithms

The maximum feasible flow in *any* network, small or large, can be found by means of a beautiful and easy-to-formulate algorithm. For ease of description, we will formulate it within the specific context of the second example of Chapter 1; see Section 1.2 and the network depicted in Figure 1.1. This will be called the *original network* in the following discussion. Its arcs will be denoted by ordered pairs of nodes; for example, the arc from J2 to D is denoted by the pair (J2, D). Consider any feasible flow on the network, i.e., nonnegative flows that do not violate arc capacities and satisfy conservation of flow at the nodes J1 and J2. A trivial choice is a flow of 0 on all arcs, which is obviously feasible, but let us consider instead, for purposes of discussion, a flow of 10 on the arcs (S, J1), (J1, J2) and (J2, D), with no flows on other arcs. (Flow units for this example are gallons per minute, and they will be not be stated explicitly from now on.) We want to improve on this *current flow*.

To do so, we construct another network associated with the current feasible flow, henceforth called the *current-residual network*. It has the *same* set of nodes as the original network, but new and potentially different arcs, along with numbers associated with arcs, are defined as follows:

- Suppose the current flow on an arc of the original network is *less* than its capacity, for example, the arc (J2, D) with current flow 10 and capacity 50. Then construct the arc (J2, D) in the current-residual network and associate with it the largest *additional* flow that is feasible on the original arc, i.e., 40.

- Suppose an arc of the original network has current flow *greater* than 0. For example, consider again the arc (J2, D) with flow 10. Then in the current-residual network, construct the arc (D, J2), i.e., from D to J2, and associate with it the largest amount by which the flow from J2 to D can be *decreased*. Obviously, this is the same as the current amount of flow from J2 to D, namely, 10. Thus, the number associated with the newly constructed current-residual network arc from D to J2 is the amount of flow that can be *negated* or reversed on the arc from J2 to D in the *original* network.

Repeat the foregoing construction for every arc in the original network that satisfies one or both of the foregoing conditions. (If both, then *two* arcs are constructed in the current-residual network as in the above example.) *Thus the newly constructed current-residual network keeps track of the amounts by which current flows on the original network can be either increased or decreased.* Note that each arc on the current-residual network has a *nonzero* number associated with it. For the foregoing choice of current feasible flows on the original network, we have the current-residual network of Figure 2.1.

Next, in the current-residual network, find *any* path from S to D with the arrows all going the same way (henceforth called a *directed path* in the network). There are several possible choices, for example, the directed path S to J1 to D; or S to J2 to J1 to D; or S to J2 to D. On a simple current-residual network, a directed path can easily be found by inspection. A mechanical procedure, or algorithm, for doing this task on a more complicated network is simple to devise: start with node S and list all the nodes in the current-residual network that can be reached by an arc directed out of S. Take all nodes on the list and from each one, add to the list all the *new* nodes that can be reached by an arc directed out of them; i.e., a node that is already on the list is not added again. Repeat the process until either the destination node D appears on the list or else no new nodes can be added to it. In the former case, the node D can be reached from S; in the

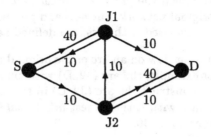

FIGURE 2.1. Current-residual network

latter, no directed path exists from S to D in the current-residual network. The foregoing mechanical approach for finding a directed path from S to D or ascertaining that none exists will be called the *forward-star procedure*.

Exercise: Show how to elaborate the foregoing procedure in order to be able to *reconstruct* the directed path from S to D. (*Hint*: Each time a new node is added to the list, preserve with it the node at the other end of the arc from whence it came.)

The directed path in the current-residual network gives a way to increase the current flow in the *original* network. Suppose we had found the directed path S to J1 to D, either by inspection or, more formally, by using the forward-star procedure. Find the *smallest* number on this path *in the current-residual network*; for our choice of path, it is 10. Then, *on the original network*, follow arcs in the sequence of nodes corresponding to this directed path, and on each arc that goes in the same direction within the directed path, *increase* the current flow by 10. On arcs in the original network that go the opposite way to the corresponding arc in the chosen directed path, *decrease* the current flow by 10. Thus, for our chosen directed path, the flow from S to J1 in the original network is increased to 20 and the flow from J1 to D is increased to 10. Suppose instead we had chosen the directed path S to J2 to J1 to D in the current-residual network. The smallest number on arcs in this path also happens to be 10. In the original network, the revised current flow from S to J2 would increase to 10, the flow from J1 to J2 would decrease to 0, and the flow from J1 to D would increase to 10. In both cases, the amount of flow being sent through the network from S to D has increased from 10 to 20 (with all flows balancing at intermediate nodes). The entire process is then repeated. It stops when no directed path can be found from S to D in the corresponding current-residual network. The current flow is then optimal; the formal verification of this intuitively evident claim is given in Section 2.3.

Let us continue the procedure and portray the sequence of current flows (on the original network) and their associated current-residual networks in Figure 2.2. The chosen directed path in each current-residual network is indicated by the sequence of arcs with two arrowheads (at the middle and end of an arc). The path is used in conjunction with the current feasible flows (shown on the original network immediately to its left) to obtain the new set of feasible flows (then shown beneath in the left-hand column). For example, in the second current-residual network of the figure, the chosen directed path is S to J1 to J2 to D. The minimum number on it is 10. The path and the feasible flows on the original network on its left are used to obtain the new feasible flows (shown on the bottom network of the left-hand column).

We now introduce some additional notation that enables us to state the algorithm formally. Let a and b denote any one of the nodes S, J1, J2, and D, and let (a, b) denote the arc from a to b (if it exists). Let $u(a, b)$ denote the capacity, or upper bound on the permissible flow, on this arc. For example, if a is J2 and b is D, then the arc (a, b) is $(J2, D)$, and its capacity $u(a, b)$ is 50. If a is J2 and b is J1, then there is no arc (a, b).

We can state an iteration of the *maximum flow algorithm* as it applies to the network of Figure 1.1:

1. Specify an initial *feasible* flow in the original network, where the flow on any arc (a, b) is either taken to be 0 or some positive *integer*, and balance of flow is maintained at all nodes. (For example, take the flow on every arc to be 0.) Denote the current flow on the arc (a, b) by $f(a, b)$.

2. Construct the current-residual network relative to the current flow on the original network as described above. Thus, on the arc (a, b) of the original network, if $f(a, b) < u(a, b)$, then construct the arc (a, b) in the current-residual network and associate with it the positive number $(u(a, b) - f(a, b))$. If $f(a, b) > 0$, then construct the arc (b, a) in the current-residual network and associate with it the positive number $f(a, b)$.

3. Find any directed path from S to D in the current-residual network using, for example, the forward-star procedure outlined above. If no directed path exists, then *stop*. The current flow on the original network is optimal. Otherwise, continue to the next step.

4. Let Δ be the smallest number on the directed path obtained at Step 3. (Note that every number associated with arcs of the current-residual

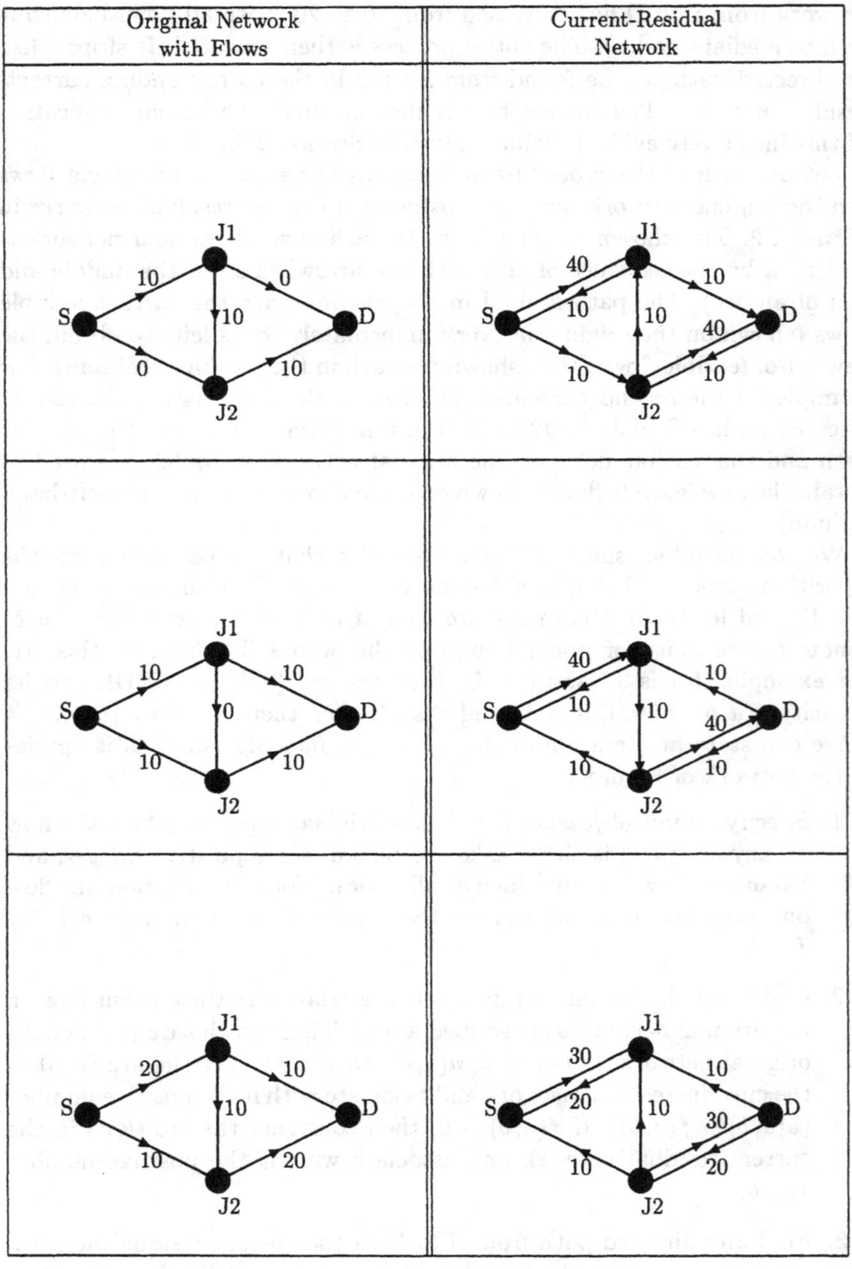

Original Network with Flows	Current-Residual Network

FIGURE 2.2. Complete example

newtork is a positive integer so $\Delta \geq 1$.) For each arc (a, b) on the directed path in the current-residual network, consider the corresponding arc in the original network from which it was derived, and its associated current flow $f(a, b)$. If the two arcs are *in the same direction* then increase the current flow on (a, b) by Δ, otherwise decrease it by Δ. Note that the total flow out of S (or equivalently into D) must then increase by Δ. Return to Step 2.

The total current flow out of S increases by at least $\Delta \geq 1$ after each iteration. Also, every arc of the original network in Figure 1.1 has a finite capacity, and thus the total current flow cannot increase without limit. The algorithm must eventually terminate at Step 3.

It is immediately evident that precisely the same algorithm can be applied to any network, not just the particular network of Figure 1.1, which was used purely for ease of description. We will continue to assume, for the moment, that every arc of the original network to which the algorithm is applied has a finite capacity given by a positive integer (whole number), and that the feasible flows on arcs used to initiate the algorithm also have integer values. Then the algorithm must terminate for reasons discussed in the previous paragraph.

Exercise: Apply the maximum flow algorithm to any network of your choice that satisfies the foregoing two assumptions used to ensure termination.

2.2.1   Refinements

Refinements of the very basic maximum flow algorithm formulated above have been explored in substantial detail in the literature. Of particular interest and relevance are the following:

- Improvements based on labeling that circumvent the need to *explicitly* generate the current-residual network at each iteration; see Kozen [1992] or Lawler [1976; 2001] for excellent descriptions and associated complexity analysis.

- The integrality assumptions on capacities and initial flows are not particularly taxing. A large artificially chosen number can be used as an upper bound on the flow on an uncapacitated arc. And, if capacities are rational numbers, the units of flow can be changed; i.e., the problem can be rescaled, so as to restore integral arc capacities. However, it is obviously preferable, from both a conceptual and a practical standpoint, to remove these assumptions from the algorithm. An elegant combinatorially-based enhancement that attains this goal was proposed by Edmonds and Karp [1972]; for details, see the two books referenced in the previous item, or see Evans and Minieka [1992; Chapter 6].

- Development of effective techniques for purposes of practical implementation, in particular, the design of appropriate data structures.

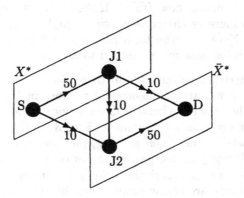

FIGURE 2.3. The optimal cutset

2.3 Duality

Consider the final current-residual network of Figure 2.2, where the node
D is not reachable by a directed path from S, resulting in termination of
the algorithm of the previous section at Step 3. Let X^* denote the set of
all nodes in this current-residual network that *can* be reached by a directed
path starting from node S, and let $\bar{X}^*$ denote the set of all other nodes.
The sets X^* and $\bar{X}^*$ are *disjoint*, together they include *all* the nodes of
the current-residual network, X^* contains S but not D, and $\bar{X}^*$ contains
D but not S. We will call this a *partition of the nodes separating S and D.*
It is defined by the set X^*, which, in turn, determines the set of remaining
nodes $\bar{X}^*$. Since the original network and the current-residual network have
the same set of nodes, the node partition just defined is obviously also a
partition of the nodes separating S and D of the *original* network.

Let $(X^*, \bar{X}^*)$ denote the set of *all* arcs leading from a node in X^* to a
node in $\bar{X}^*$ in *the original network*. This will be called the *cutset* defined
by X^*. Similarly, the set of all arcs from a node in $\bar{X}^*$ to a node in X^* will
be denoted by $(\bar{X}^*, X^*)$ and called the *backward cutset* defined by $\bar{X}^*$.

The partition of nodes defined by X^* and the associated cutset $(X^*, \bar{X}^*)$
obtained upon termination of the above algorithm are depicted in Figure
2.3. (Note that the node partition is determined from the current-residual
network, but it is depicted on the *original network*. The cutset is determined
from the original network.) Arcs with double arrowheads correspond to the
cutset $(X^*, \bar{X}^*)$. The set of arcs $(\bar{X}^*, X^*)$, for the backward cutset defined
by $\bar{X}^*$, is empty for this particular example, but this will not be true in
general.

The sum of the capacities of all arcs in $(X^*, \bar{X}^*)$ will be called the *capacity of the cutset* and denoted by $u(X^*, \bar{X}^*)$; and the sum of the current flows on all arcs in $(X^*, \bar{X}^*)$ will be denoted by $f(X^*, \bar{X}^*)$. (Analogous quantities can be defined for the backward cutset.) Now the current feasible flow on every arc of $(X^*, \bar{X}^*)$ must be *at its capacity*, because otherwise, an arc leading from a node in X^* to a node in $\bar{X}^*$ would have been constructed in the current-residual network; and the set $\bar{X}^*$ would then contain a node reachable by a directed path from S, which is a contradiction. Thus, $f(X^*, \bar{X}^*) = u(X^*, \bar{X}^*)$. (As we have seen above, the backward cutset $(\bar{X}^*, X^*)$ is empty. If it were not, as can happen on other network examples, then the flow on all arcs in the backward cutset must be 0. Otherwise, an arc would have been constructed in the current-residual network from some node in X^* to a node in $\bar{X}^*$, again in contradiction of the fact that X^* is as large as possible. Thus, at the termination of the algorithm on *any* example, $f(\bar{X}^*, X^*) = 0$.)

Picture two "conduits" joining the sets X^* and $\bar{X}^*$. The first is composed of all arcs in $(X^*, \bar{X}^*)$, and with its direction of flow the same as for arcs in this set, i.e., from X^* to $\bar{X}^*$; the second is composed of arcs in $(\bar{X}^*, X^*)$, with direction of flow from $\bar{X}^*$ to X^*. The total current flow in the first "conduit" is $f(X^*, \bar{X}^*)$, and in the second it is $f(\bar{X}^*, X^*)$. The balance of flow in the network then clearly implies that

Total flow out of S = Total flow into D = $f(X^*, \bar{X}^*) - f(\bar{X}^*, X^*)$.

Using the observation that flow in the cutset is capacitated and flow in the backward cutset is zero, it follows that

Total flow out of S = Total flow into D = $u(X^*, \bar{X}^*)$.

The algorithm has terminated with a current total flow that equals the capacity of the associated cutset $(X^*, \bar{X}^*)$. We assert that this is the maximal flow that can be sent through the network. To validate this assertion, let us consider *any* partition of the nodes of the original network into *disjoint* sets X and $\bar{X}$ separating S and D; i.e., together X and $\bar{X}$ include all nodes of the network, S is in X, and D is in $\bar{X}$. Define the cutset $(X, \bar{X})$, its capacity $u(X, \bar{X})$, the backward cutset $(\bar{X}, X)$, and the flows $f(X, \bar{X})$ and $f(\bar{X}, X)$ in an analogous way to the quantities defined at the termination of the algorithm for X^* and $\bar{X}^*$. Again balance of flow in the network implies that

Total flow out of S = Total flow into D = $f(X, \bar{X}) - f(\bar{X}, X)$.

Since the flow on each arc is feasible, it is obviously at most the capacity of that arc. Also, the flow on every arc is nonnegative. Then the foregoing statement immediately implies that

Total flow out of S $\leq u(X, \bar{X})$ = Capacity of the cutset $(X, \bar{X})$.

Thus a total feasible flow out of S (or into D) in the network cannot exceed the capacity of *any* cutset, and we know that the algorithm terminated with a flow that was *equal* to the capacity of a particular cutset identified by the algorithm. It follows that the flow at termination must be maximal and the cutset $(X^*, \bar{X}^*)$ must have minimal capacity vis-à-vis any other cutset $(X, \bar{X})$.

The foregoing statements are illustrated in Figure 2.4, which depicts all possible node partitions X and $\bar{X}$ for the illustrative network of Figure 1.1, along with the corresponding cutset $(X, \bar{X})$. (Note that the definition of a partition of nodes separating S and D, stated at the beginning of the present section, implies that there are *only four* distinct partitions. For each partition, arcs with double arrowheads define the associated cutset.) The cutset with the least capacity is 30. This corresponds to the node partition and cutset identified within the final current-residual network of Figure 2.2 when the maximum flow algorithm terminates. The current total flow out of S at termination is 30, which equals the minimum capacity cutset, and this flow is maximal. (Note that in general, there may be alternative ways of achieving the maximum total flow.)

We can summarize the foregoing observations by the following statements, which indeed apply to any max-flow problem network:

- A feasible (total) flow that can be sent through a given network, from a source node to a destination node, *cannot exceed* the capacity of *any* cutset of the network. This is called the *weak duality* principle for the maximum flow problem.

- The maximum total flow that can be sent through a given network, from a source node to a destination node, *equals* the minimum capacity cutset of the network. This is called the *strong duality* principle for the maximum flow problem (also known as the *max-flow/min-cut theorem*).

- The capacity of the cutset identified at termination of the maximum flow algorithm is smaller than or equal to the capacity of *every* other cutset defined on the network.

- The maximum flow algorithm will terminate provided there is at least one node partition whose associated cutset has *finite* capacity. Thus one can remove the earlier restriction that every arc on the network must have a finite capacity. (We retain the integrality assumption on capacities and on the feasible flow used to initiate the algorithm. The removal of this assumption was touched on in Section 2.2.1.)

Duality enables us to solve a broader set of problems, for example, the following: let us interpret the network of Figure 1.1 as a set of one-way

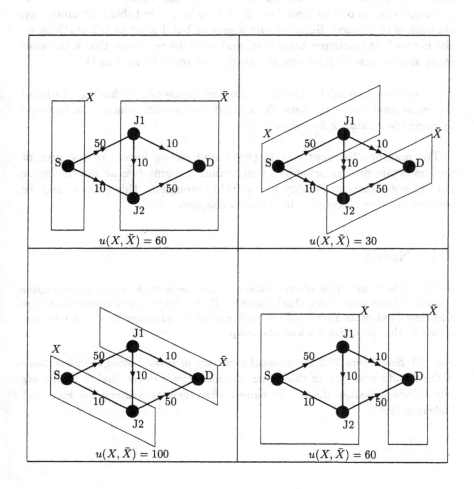

FIGURE 2.4. All possible cutsets

roads, and the number associated with an arc as the cost of building a toll station on the road to which that arc corresponds. Thus, the cost of building a toll station on the road from S to J1 is 50 (in suitable cost units, say, thousands of dollars). Suppose one wants to build a set of toll stations on the network at minimum total cost, and in so doing ensure that a traveller must always pass at least one toll station en route from S to D.

Exercise: Show that solving the above problem corresponds to finding a cutset of minimum total cost in the network and that this optimal solution can be found by using the maximum flow algorithm.

The max-flow/min-cut results (weak and strong forms) of the present section typify duality principles that manifest themselves across the entire field of optimization. Duality is a refrain throughout this primer, and we resume our discussion of it in the next chapter.

2.4   Notes

Sec. 2.0: The itemized list of network application areas at the start of this chapter is derived from Rockafellar [1998, Chapter 1]. For an in-depth treatment of the max-flow (and other network-flow) problems at an advanced level, we refer the reader to this definitive work on the subject.

Sec. 2.3: See Figure 10.2 for a visual summary of several other manifestations of the duality principle. In this figure, the max-flow/min-cut duality results are called Ford–Fulkerson duality, so named after their originators; see Ford and Fulkerson [1962].

3
Duality on Bipartite Networks

Duality was introduced in the previous chapter within the setting of the max-flow optimization problem and its relationship, Janus-faced, with the min-cut optimization problem. The underlying *duality principle* postulates, in general, the structural correspondence between a given optimization problem, called the *primal*, and an associated optimization problem, called the *dual*. The solution of the one yields important and often complete information about the solution of the other.[1]

As Rockafellar [1998] has noted, in a statement made specifically with reference to the subject of network flows, but applicable to the field of optimization across the board:

> Duality appears at all levels and dominates much of the subject. It serves to draw attention to many aspects of parallelism between different parts of the theory, thereby simplifying ideas and frequently suggesting how an approach in one context may be carried over to another. It leads to computational techniques that ... are often highly effective.

Although the full significance of the foregoing quotation can be appreciated only after an in-depth study of optimization, it nevertheless serves to highlight the fact that the duality principle is one of the great mathematical ideas of the subject, and *arguably its greatest!* Thus it will come as no surprise that duality will be a repeated refrain throughout this primer.

---

[1]The primal is usually stated in terms of minimizing its objective and the dual in terms of maximizing its objective, but the two designations are interchangeable.

In the present chapter, we will discuss the duality principle within a second network-based setting—optimization problems of *matching* and *covering* on a particular class of networks termed *bipartite*—that builds very directly on the max-flow/min-cut duality results of Chapter 2.

The duality principle can also be introduced, and intuition thereby enhanced, within a simple setting of geometry. Consider a flat circular disk in two dimensions, say $\mathcal{P}$, a point z outside the disk, and the tangents to the disk's boundary that pass through z. Let A and B be the points where these two tangents touch the disk, as shown in Figure 3.1.

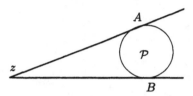

FIGURE 3.1. Duality in geometry

Let l be any line that forms a tangent to the disk at a point on its boundary that lies to the left of A and/or B, and denote the set of all such lines by $\mathcal{L}$. Let the distance from z to a *point* p on the disk be defined in the usual way as the length of the segment joining the two points; and let the distance from z to a *line* l in $\mathcal{L}$ be defined to be the *length of the perpendicular segment* dropped from z to the line l.

Exercise: In Figure 3.1, draw a particular line l as described above and the perpendicular segment defining the distance of z to l.

Exercise: Define the *primal* to be the optimization problem of finding the point p^* that lies on the disk $\mathcal{P}$ and is as *close* as possible to z. (Its objective function to be minimized is the distance between z and any point p that is constrained to lie on the disk.) Define the *dual* to be the optimization problem of finding the line l^* for which the distance from z to any line of $\mathcal{L}$ is a *large* as possible. Verify geometrically that the relationship between the foregoing primal and dual problems is completely analogous to the relationship between the max-flow and min-cut problems of Chapter 2, and that the two optimization problems defined on Figure 3.1 exhibit weak and strong duality results analogous to the ones described in Section 2.3.

	1	2	3	4	5	6	7
1		C		C	C		C
2					C		C
3		C			C		
4	C	C	C		C	C	
5		C					C
6					C		
7		C			C		C

	1	2	3	4	5	6	7
1		1		C	C		C
2					1		C
3		C			C		
4	1	C	C		C	C	
5		C					1
6					C		
7		C			C		C

TABLE 3.1. The Hollywood problem

3.1 Matching

Let us consider the *Hollywood problem*[2] of finding seven brides for seven brothers. Not all the candidate brides are compatible with all the seven brothers, their prospective grooms. Compatibilities are shown in the upper table of Table 3.1; the *compatibility matrix*. The brothers, or grooms, correspond to the rows, and the brides correspond to the columns; a symbol C indicates compatibility, and a "blank" indicates incompatibility. Thus, for example, groom 2 and bride 5 are compatible with one another; groom 6 and bride 3 are not. Bride 5 is obviously very popular, as is groom 4.

Let us match grooms to brides in the obvious way: match brother 1 with the first compatible bride, then brother 2 with the first compatible bride that is still available, and so on. If there is no available bride that is compatible with the brother currently being matched, then continue on to the next brother. When we highlight each matching obtained in this way by replacing the C with a boldface 1, we obtain the matchings (marriages) shown in the second table of Table 3.1. The cells of the compatibility matrix containing 1's are said to be *independent cells*, because each 1 occurs in a unique row and column of the matrix, i.e., no row or column contains more than a single 1. Only four brothers can be married with this simple selection

[2]So called after the Cinemascope movie classic *Seven Brides for Seven Brothers*, directed by Stanley Donen (1954).

procedure. Can we do better? It is easy to see that five brothers can be married if we start by pairing brother 1 with bride 4 and then continue with the straightforward matching procedure. We can find 5 independent cells by this procedure. Can one improve further on this matching, i.e., what is the *largest number of independent cells of the compatibility matrix that can be identified?*

To address this question, let us represent the problem in a different way, as shown in the uppermost network of Figure 3.2, formally called a *bipartite network*. (Because arrowheads are not associated with arcs, the network is called *undirected*, in contrast to Figure 1.1, where the arcs and the network are *directed*. The undirected network is formally called an undirected *graph*, but we will not use the latter terminology here.) The seven nodes on the left-hand side correspond to the brothers, and the seven on the right correspond to the brides. The brother, or groom, nodes are labeled l1, ..., l7, and the bride nodes are labeled r1, ..., r7, where "l" means "node on the left" and "r" means "node on the right." If a brother is compatible with a bride, then an arc joins the corresponding nodes. The matching of four brides to grooms using the straightforward procedure of the preceding paragraph is shown in the second copy of the network, with the four arcs defining the matching now highlighted using thick lines. (The bottom copy of the network in this figure will be discussed in the next section.)

We can find the largest matching in the bipartite network by formulating the problem as a maximum flow problem as shown in the upper half of Figure 3.3. An artificial (source) node s is added on the left and connected to all groom nodes, and another artificial (terminal) node t is added on the right and connected to all bride nodes. The flow on each arc is now directed from left to right, but for convenience, arrowheads are not shown explicitly. A capacity of 1 is specified on *every* arc of the newly constructed network.

The optimal matching can then be found by applying the maximum flow algorithm of Section 2.2 to this network, and in this way it can be verified that the largest flow, and hence the largest number of brides that can be matched to brothers, is 5. To do so, let us assume we begin with the feasible solution stated above that can match 5 brides to grooms. In the upper network of Figure 3.3, an arc with thick lines has a current (feasible) flow of 1, and the flow on all other arcs (thin lines) is zero. The associated current-residual network is given in the lower half of Figure 3.3. The terminal node t cannot be reached from s, and thus the current feasible solution must be optimal. (Note that the optimal solution itself is not unique; i.e., there are other ways to achieve the maximum possible flow of 5.) The associated optimal cutset $(X^*, \bar{X}^*)$ is shown in Figure 3.4, where the original nodes are regrouped into the sets X^* and $\bar{X}^*$ that define the optimal node partition (identified from the residual network) and then depicted as such in the figure.

Exercise: Add one additional compatibility entry in position $(7, 1)$ of Table 3.1 and introduce an additional arc from l7 to r1 in the upper network of Figure 3.3 with current flow 0. Modify the current residual network and show that the current flow in the network is no longer optimal. Continue the maximal flow algorithm to find the maximum number of matchings (marriages) that are now possible.

The terms "matching" and "assignment" are often used interchangeably in the foregoing network optimization problem. Also, one can easily give alternative, more practical, interpretations of the data used. For example, for "brides" substitute "job applicants" or "umpires," and for grooms substitute "jobs" or "baseball games."

3.2   Covering

Let us consider another interpretation of the data of the previous section. Suppose the uppermost network of Figure 3.2 depicts the transatlantic flights of an international airline, with nodes on the left representing seven European cities and nodes on the right representing seven American cities. An arc joining a pair of nodes denotes an airline route (two-way) between the corresponding two cities. In order to enhance employee morale, the airline's management has decided to establish "sister relationships" between its staff at European and American cities. Sister cities must have an existing airline route joining them. No city can have more than one sister city, and management wishes to create as many sister relationships as possible. This can be done by solving the matching problem on the network, and its optimal solution is obviously identical to that of the Hollywood problem i.e., at most five sister-city relationships can be established.

Now consider *a different optimization problem defined on the same data* as follows: a joint air-transportation agreement permits the airline to set up customs inspection stations at its European and American city terminals, and passengers travelling on the airline can be processed at either the start or end of their journey. The airline wishes to construct as *few* customs stations as needed to ensure that a passenger travelling on any existing route passes through a city with an inspection station. (If the nodes on both ends of a route have an inspection station, then the airline can process passengers at either end, as suits its convenience.) So called *covering* problems of this type are considered here; the reader will very likely have guessed that they are *dual* to the matching problems of the previous section.

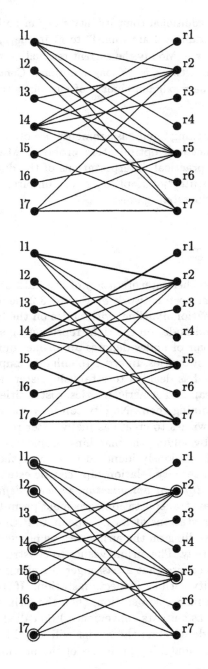

FIGURE 3.2: Bipartite network, a matching, and a cover

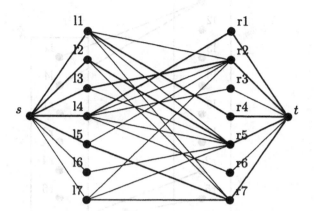

For convenience, arrowheads are not depicted explicitly. *All arcs above are assumed to be directed from left to right.* The capacity on every arc is 1. An arc with a thick line has a current (feasible) flow of 1. All other arcs have flow 0.

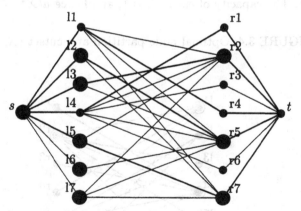

Again, for convenience, arrowheads are omitted. Arcs with thick lines are *directed from right to left.* All other arcs (with thin lines) are directed from left to right. Every arc (implicitly) has the number 1 attached; i.e., it can carry an additional (or negate a previous) flow of 1. Nodes that are larger in size can be reached by a *directed* path from s, e.g., l5 via l7 and r7. In constrast, l1 or r1 cannot be reached by such a path.

FIGURE 3.3: Max-flow and current-residual networks

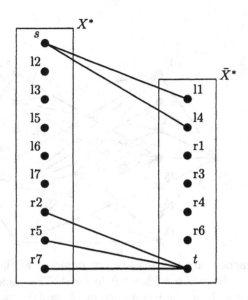

Only arcs that define the cutset are depicted, i.e., arcs from a node in X^* to a node in $\bar{X}^*$. Other arcs, e.g., from l3 to r2, or from l1 to r2, are not shown. The capacity of each arc is 1, and hence $u(X^*, \bar{X}^*) = 5$.

FIGURE 3.4: Optimal node partition and cutset $(X^*, \bar{X}^*)$

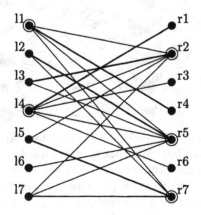

At least one end node of *every* arc is circled (contained in the optimal cover: l1, l4, r2, r5, and r7). The thick-line arcs identify the associated optimal matching as derived from Figure 3.3.

FIGURE 3.5: Optimal cover and matching

Formally, a *cover* of the nodes of the network is defined to be a subset of the entire set of nodes such that *every* arc of the network is incident on (goes out of or goes into) *at least one* node in the cover. An example of a cover is given by the circled nodes in the bottom network of Figure 3.2. In terms of the matrix of matchings in Figure 3.2, this corresponds to drawing vertical lines through the C's in columns 2 and 5 and horizontal lines through the C's in rows 1, 2, 4, 5, and 7, i.e., covering entries in the rows and columns of the compatibility matrix in Table 3.1 with lines so that all the C's in the matrix are covered. What is the cover of the nodes of the bipartite network that contains the *smallest* number of nodes? In other words, what is the smallest number of lines through rows and columns that can cover all the C's in the compatibility matrix? The answer is intimately related to the solution of the matching problem of the previous subsection, as we will see next.

3.3   König–Egerváry Duality

The celebrated König–Egerváry duality theorem (König [1936], Egerváry [1931]), one of the most beautiful results of network optimization, states that the *maximum number of arcs in a matching* on a bipartite network *equals* the *minimum number of nodes in a cover* of the network. Stated in another way and now with specific reference to the matching/covering examples of Sections 3.1–3.2 and the data in Table 3.1; the largest number of *independent cells* in the compatibility matrix of Table 3.1—recall from Section 3.1 that this is the same as the maximum matching in the associated network—equals the smallest number of *lines* across its rows and/or columns that are needed to cover all the C entries of the matrix. Verify that this number is the same as a minimum cover of the network. Therefore the solution of the covering problem of Section 3.2 can be found from the solution of the matching problem of Section 3.1. The two problems are dual to one another.

We can explore this further by returning to the optimal solution of the matching problem of Table 3.1 and its interpretation in terms of airline routes given at the beginning of Section 3.2. The associated optimal cutset is shown in Figure 3.4.

Now consider the following rule for choosing nodes of a set C of the uppermost network of Figure 3.2 by using the optimal partition of its nodes given in Figure 3.4 (ignore the nodes s and t):

- If a node from the right-hand side of the network—an American city in the airline interpretation of the data—is in X^*, then include it in the set C.

- If a node from the left-hand side of the network—a European city in the airline interpretation—is in the complementary partition $\bar{X}^*$, then include it in the set C.

The five nodes in C identified by the above rule correspond to the circled nodes in Figure 3.5, and one can verify that C is indeed a cover of the network. Furthermore, no other cover of the network can contain *fewer* than 5 nodes. This is because we know that an optimal matching contains 5 arcs (Section 3.1), and associated with these arcs is a set of 5 *distinct pairs* of nodes of the network. If one seeks nodes that cover just these 5 arcs in the optimal matching—they are obviously arcs of the original network—then one would have to select a node from each pair. Arcs not in the optimal matching could conceivably require that more nodes be added in order to obtain a cover of the entire network. In other words, *at least* 5 nodes are needed in any cover of the network. But the foregoing rule has identified a cover that contains exactly 5 nodes, and thus the cover C as depicted in Figure 3.5 must be an *optimal cover*. (Note again that just as there are alternative ways to achieve a maximal matching of 5, there are also alternative ways to achieve a minimal cover of 5.)

Our selection rule for finding this cover has worked for our particular example! Of course, its validity must be established *more formally and in general*. We will not spell out all the details, but they are well within the grasp of a motivated reader who sets out to follow the trail leading from Figures 3.2 to 3.5, and their associated max-flow/min-cut duality discussed in the previous chapter. Alternatively, the reader can devise and work through a few more examples in order to develop additional empirical evidence for the validity of König–Egerváry duality.

Exercise: Justify the rule used to find the foregoing optimal cover (i) for the example, and (ii) in general.

3.4 Notes

Sec. 3.3: A useful reference for the results of this section is the elegant book of Lawler [1976; 2001].

4

A Network Flow Overview

We have now encountered two key optimization models defined on networks: maximum flows and matching. We have also encountered their respective dual models: minimum cuts and covering. Our aim in the present chapter is to give a broad overview of several other key network flow optimization models and their interrelationships, and to summarize them in the form of a schematic tree. This portrayal, in turn, enables us to highlight and constrast the two main perspectives—combinatorial vis-à-vis continuous—on network flow optimization.

4.1 Problem Reformulations

Let us return to the maximum flow problem of Chapters 1 and 2 and consider key reformulations of this problem that open a window on other network optimization problems.

4.1.1 Optimal Cost/Profit Network Flow Model

In the network shown in Figure 1.1 and the associated maximum flow problem that was discussed, in detail, in Chapter 2, a feasible flow on any arc (a, b) was denoted by $f(a, b)$. Additionally, let F denote the total flow out of the source node S (or into the destination node D). Let us introduce an *artificial* arc that sends the flow F out of the node D and back into the node S, i.e., $f(D, S) = F$, and place an infinite capacity on this arc (denoted

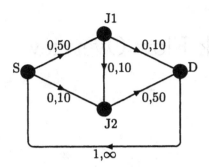

FIGURE 4.1. Maximum profit network flow reformulation

by ∞). Also introduce "unit profits" into the maximum flow model as follows: let the profit from sending one unit of flow through the artificial arc be +1 and the profit from sending a unit of flow through any other arc of the original network be 0. Let the capacities on the original arcs be as before. This reformulation is depicted in Figure 4.1, where two numbers are now associated with each arc, namely, the profit and the capacity. Then it is evident that the problem of maximizing the total profit derived from flows in the network is the same as maximizing the flow that can be sent through the network. This reformulation is an example of an *optimal cost/profit network flow model.*

4.1.2 An Equivalent Linear Programming Formulation

For the network of Figure 4.1, the flow-balance constraint at the source node S can be expressed as follows:

$$f(D,S) = f(S,J1) + f(S,J2), \quad \text{or} \quad -f(D,S) + f(S,J1) + f(S,J2) = 0,$$

where the above quantities were introduced in the previous subsection.

Flow-balance constraints can be written at each of the other nodes, for example, at J1:

$$f(S,J1) = f(J1,D) + f(J1,J2), \quad \text{or} \quad -f(S,J1) + f(J1,D) + f(J1,J2) = 0.$$

Then one can express the maximal flow problem, or its maximum profit network flow equivalent given in the previous subsection, as the problem of maximizing the variable F, or $f(D,S)$, subject to flow-balance constraints at all nodes. In addition, the flow on each arc must be feasible; i.e., it must not exceed the capacity of the arc and it must be nonnegative, for example, $f(S,J1) \le 50$ and $f(S,J1) \ge 0$, or equivalently, $0 \le f(S,J1) \le 50$. This is the statement of the maximum flow problem (or its max-profit form of the previous subsection) reexpressed explicitly as a *linear program.*

Exercise: Write out the complete linear program outlined above for the maximum flow problem of Figure 1.1 or Figure 4.1. State each flow-balance equation so the right-hand side is zero, and then verify that each variable of the linear program, for example $f(J1, D)$, occurs in just two flow-balance equations, once with coefficient $+1$ and once with coefficient -1.

Exercise: Extend the network of Figure 1.1 to a larger network of your choice, and again verify that each column of the associated explicit linear program has a single $+1$ entry and a single -1 entry, with all other entries being 0.

As highlighted by the foregoing discussion and exercises, the explicit linear program associated with a maximum flow network problem with many nodes and arcs can be very large, but it is simultaneously very *sparse*. Thus linear programs arising from the max-flow problem (and other network-flow problems considered later in this chapter) provide an excellent illustration of *characteristics of practical linear programs that hold in general*; namely, they are large-scale, but they also have relatively few nonzero elements. Consider, for example, max-flow networks with side constraints on the flows. (Specfically, in Figure 4.1, add the constraint that the total flow out of node J1 cannot exceed the total flow out of node J2.) On the equivalent linear program, the *pure network property* is lost; i.e., it is no longer true that each column of the LP matrix contains just a single $+1$ element and a single -1 element; but nevertheless, each column will have very few nonzero elements. Other examples of large, sparse linear programs will be encountered in subsequent chapters, in particular, in Chapter 9.

4.2   A Network Flow Tree

We encountered the *maximum flow* model in Chapters 1 and 2 and observed, in Section 4.1, that it can be formulated as a particular instance of the optimal cost/profit network flow model. And the latter, in turn, can be formulated as a linear programming model.

In full generality, the *optimal cost/profit network flow* model has unit costs or profits associated with all arcs, together with capacities on flows. In addition, supplies or demands can be placed at nodes of the network. For example, in Figure 4.1, one can have an additional water source at J1 and a demand for water at J2. Models of this type can also be expressed explicitly as linear programs with pure network structure, in a manner analogous to the reformulation in Section 4.1.2.

The bipartite *matching* model was encountered in Chapter 3. It can be reformulated as a max-flow model of a special type, as discussed in Section 3.1, and thus it is evident that a bipartite matching model can also be converted into an optimal profit network flow model and a linear programming model.

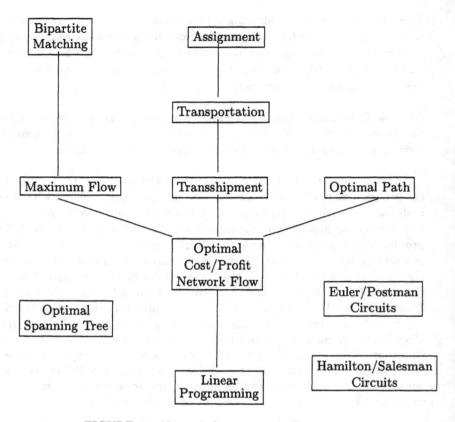

FIGURE 4.2. Network flow tree and offshoots

The foregoing interrelationships are depicted in the the left-hand part of the schematic of Figure 4.2, which takes the form of a "tree." A network flow model higher up in the tree can be formulated as a particular instance of a model lower down whenever they are joined by a path in the tree. For example, a path links the matching model and the optimal cost/profit network flow model. The "root" of the tree is the *linear programming* model.

The central and right-hand branches in the tree schematic depict (a selected set of) other key network flow models.

Transportation and *transshipment* models are particular types of optimal cost or optimal profit network flow problems defined on more restricted networks. A transportation model is defined on a bipartite network—see Chapter 3—with a set of "source" nodes (by convention on the left of the network) and a set of "destination" nodes (on the right). Directed arcs in the bipartite network join sources and destinations, with the direction of flow always being from source to destination. Transshipment models allow intermediate, or transshipment, nodes between the sources and destinations, and again directed arcs always point in the same direction, say from left to right. Applications for models of these types abound and often involve the transportation or transshipment of goods, products, and services from suppliers to consumers.

The *assignment* model, at the top of the tree, is a counterpart of the bipartite matching problem in which a number is associated with an arc; i.e., a cost is incurred or a profit is derived when two nodes in the network are matched, or assigned, to one another. For example, matching a person to a job may incur a cost. The aim in the assignment problem is then to find a matching, or assignment, at optimal total cost/profit. The assignment model can be reformulated as a specialized form of the transportation model.

Problems of finding an optimal directed path were encountered in Chapter 2 as subproblems within the max-flow algorithm. Such problems are also of interest in themselves, and they arise in many practical applications. The *shortest-path* problem, which can be defined in directed or undirected networks, is simply the problem of finding the directed (or undirected) path between two particular nodes for which the sum of numbers on arcs is *no larger* than that for any other path between the two nodes. It can be expressed as a particular type of optimal cost/profit network flow problem. (Try this as an exercise!) Longest-path problems are defined analogously with "no larger" replaced by "no smaller."

Finally, three models are depicted as "offshoots" in Figure 4.2. Optimization models involving a "tree network structure" arise as subproblems within the formulation of the network simplex algorithm, as we shall see later, in Chapter 6, and they are also important for practical applications in their own right Interestingly enough, usage of the term "tree" in these models is generically the same as its usage within the tree schematic of Figure 4.2. Formally, a *tree* is an undirected network that has the following two properties: (i) a path exists between every pair of nodes; (ii) the network contains no cycles, i.e., no paths that start and end at the same node.

Consider an arbitrary undirected network. A *spanning tree* of this network is obtained by taking *all* the nodes of the network and a *subset* of its arcs in such a way that the resulting *sub*network is a tree, i.e., satisfies the foregoing two properties. A network will usually contain many spanning trees. When costs/profits are associated with its arcs, an *optimal spanning tree* for the network is a particular spanning tree for which the total cost/profit, i.e., the sum of the numbers on the arcs of the tree, is optimal (smallest/largest). There are many practical applications for such models. Think, for example, of an indirected network in which nodes represent junctions, arcs represent roads that could be built between nodes, and the numbers on arcs represent the estimated costs of construction. The aim is to select the set of roads (arcs) to physically construct so that a person can travel by road from each node to every other node, and the total construction cost of roads built is minimized. For a second example, replace "roads" by "telephone lines."

An *Eulerian circuit* problem involves finding a path in a network that starts and ends at the same node and visits all arcs of the network once and only once en route. Replace the phrase "arcs of the network" by "nodes of the network" to define the *Hamiltonian circuit* problem. *Postman* and *traveling salesman* problems involve networks with costs (or profits) on arcs for which minimum cost (or maximum profit) Eulerian and Hamiltonian circuits, respectively, are sought. Again applications abound: optimal routing of delivery vehicles, snow plow routing, the optimal route for mowing roadside verges, and so on. Such models are not explicitly part of the tree schematic in Figure 4.2, but algorithms for solving them rely extensively on the solution of appropriately defined *sequences of subproblems* defined from models that lie within the tree schematic.

Figure 4.2 depicts primal network flow models, and each of the models within the tree schematic has a dual. For example, we have encountered the dual counterparts of the bipartite matching and maximum flow problems in earlier chapters, namely, bipartite covering and minimum cuts.

4.3   Summary: Combinatorial vis-à-vis Continuous

The tree schematic of Figure 4.2 is by no means exhaustive. It depicts a *representative set* of network flow optimization models that frequently arise in practice, and their interrelationships. Algorithms for solving them can be obtained in two main ways.

They can be devised in piecemeal fashion for each of the models, often independently of relationships to other models in the schematic. This development *lacks an overall unity*, but it has produced a very rich variety of beautiful and efficient algorithms. We have seen examples of this so-called combinatorial, or "top-down," approach in Chapters 2 and 3. A comprehensive discussion of network flow algorithms for solving models of Figure 4.2 (and many others) from this perspective can be found in the classic textbook of Lawler [1976, 2001]. For a more accessible and very readable account, see Evans and Minieka [1992], which also provides easy-to-use software for experimentation.

The alternative approach is to start at the linear programming root of the tree and develop *specializations* of key linear programming algorithms—simplex and interior-point algorithms to be discussed in Chapters 6 and 7—that take into consideration the *detailed structure of the linear programming equivalent* of the network flow model being addressed. For example, the network simplex algorithm for solving the optimal cost/profit network flow problem is obtained in this way, as will be outlined at the end of Chapter 6. Specializing the simplex algorithm gives the network flow development much greater algorithmic unity, but much less algorithmic variety. For a comprehensive and unified discussion of this so-called continuous, or "bottom-up," approach, see the textbook of Bazaraa et al. [1990].

Chvátal [1983] gives an especially elegant presentation of the subject of linear programming and network flows, and strikes a good balance between the combinatorial and continuous perspectives.

4.4 Notes

Sec. 4.2: Lawler [1976; 2001] is a useful reference for the many interrelationships between network flow models; see, in particular, Chapter 5 of his book. Rockafellar [1998] gives a definitive and unified (via duality) treatment of network flow models and algorithms at an advanced level.

The pioneering work on network flow optimization is the monograph of Ford and Fulkerson [1962]. A *comprehensive list* of combinatorial optimization and other problems that are defined on networks and graphs can be found in the appendix of Garey and Johnson [1979].

5

Duality in Linear Programming

In earlier chapters, we encountered linear programming problems that were derived from underlying network flow models (see Section 4.1.2) and noted that their basic structure involves maximizing or minimizing a linear objective function, subject to a set of linear equality and/or inequality constraints, and a set of upper and/or lower bounds on the variables.

In the present chapter, we turn to the linear programming model in its own right, beginning with the duality theory that is its centerpiece. Then, in the two subsequent chapters, we describe simplex and interior-point algorithms for solving linear programs.

5.1   In the Marketplace

Let us return to the first example of this book, shopping for seafood as discussed in Section 1.1. But now, instead of buying only seafood, let us broaden the example to permit a wider variety of food purchases and consider the following problem: a budget-conscious consumer wishes to purchase, at minimum cost, suitable quantities of three basic foods at the supermarket, say, poultry, spinach, and potatoes, so that his or her daily diet provides at least 65 grams of protein, 90 grams of carbohydrate (energy), 200 milligrams of calcium, 10 milligrams of iron, and 5000 international units of vitamin A. The nutritive food value contained in 100 grams of each of the three basic foods and its cost are summarized in Table 5.1.

Suppose the consumer purchases x_1 units (a unit is taken to be 100 grams) of poultry, x_2 units of spinach, and x_3 units of potatoes. Then his

	Poultry	Spinach	Potatoes
Cost (cents)	40	15	10
Protein (g)	20	3	2
Carbohydrate (g)	0	3	18
Calcium (mg)	8	83	7
Iron (mg)	1.4	2	0.6
Vitamin A (I.U.)	80	7300	0

TABLE 5.1. Nutrition information

or her problem can be stated as follows:

$$
\begin{aligned}
\text{minimize } z = \quad & 40x_1 + 15x_2 + 10x_3 \\
\text{s.t.} \quad & 20x_1 + 3x_2 + 2x_3 \geq 65, \\
& \qquad\;\; 3x_2 + 18x_3 \geq 90, \\
& 8x_1 + 83x_2 + 7x_3 \geq 200, \\
& 1.4x_1 + 2x_2 + 0.6x_3 \geq 10, \\
& 80x_1 + 7300x_2 \qquad\;\; \geq 5000,
\end{aligned}
$$

$$x_1 \geq 0, \; x_2 \geq 0, \; x_3 \geq 0.$$

In another part of the supermarket is a food-supplements counter, where one can purchase the nutrients that one needs directly, in the form of pills or powders. The supplier of nutrients must decide on a suitable *price* to charge for the five different nutrients desired by the consumer. Since these are the unknowns or decision variables for the supplier, let us denote these prices by y_1 cents/gram of protein, y_2 cents/gram of carbohydrate, y_3 cents/mg of calcium, y_4 cents/mg of iron, and y_5 cents/I.U. of vitamin A. Thus the income to the supplier, when the consumer purchases nutrients in pill (or powder) form that are equivalent to what could be obtained from the purchase of 100 grams of poultry, is

$$20y_1 + 0y_2 + 8y_3 + 1.4y_4 + 80y_5.$$

Note that the coefficients come from the first column of the data in Table 5.1. The prices charged by the supplier must be fair prices; i.e., the amount that the supplier charges for the equivalent of 100 grams of poultry should not exceed the amount that the consumer would have to pay for 100 grams of poultry itself. Otherwise, the consumer would have no financial incentive for the direct purchase of nutrients. Therefore, the supplier will want to satisfy the constraint

$$20y_1 + 0y_2 + 8y_3 + 1.4y_4 + 80y_5 \leq 40.$$

Similar constraints can be formulated for the other two foods, spinach and potatoes.

As stated above, the consumer needs 65 grams of protein, 90 grams of carbohydrate, 200 milligrams of calcium, 10 milligrams of iron, and 5000 I.U. of vitamin A. If the supplier provided them directly to the consumer, then the income to the supplier would be

$$65y_1 + 90y_2 + 200y_3 + 10y_4 + 5000y_5.$$

The supplier wants to maximize his income subject to setting fair prices as constrained above. Thus the supplier would formulate a linear program as follows:

$$
\begin{array}{rl}
\text{maximize } w = & 65y_1 \; + \; 90y_2 \; + \; 200y_3 \; + \; 10y_4 \; + \; 5000y_5 \\
\text{s.t.} & 20y_1 \qquad\qquad\quad + \; 8y_3 \; + \; 1.4y_4 \; + \; 80y_5 \;\le\; 40, \\
& 3y_1 \; + \; 3y_2 \; + \; 83y_3 \; + \; 2y_4 \; + \; 7300y_5 \;\le\; 15, \\
& 2y_1 \; + \; 18y_2 \; + \; 7y_3 \; + \; 0.6y_4 \qquad\qquad\;\le\; 10, \\
& y_i \ge 0, \quad i = 1,\ldots,5.
\end{array}
$$

Observe that the two linear programs are defined in terms of the same data in Table 5.1, the so-called *primal* from the original data, and the *dual* from its transposition.

As one might expect, the cost to the consumer for any feasible purchase of food is *no smaller* than the income to the supplier for any feasible prices that are set by the latter, i.e., $z \ge w$ for associated feasible values of the variables $x_1, \ldots, x_3$ and $y_1, \ldots, y_5$. This is the *weak duality* result for the foregoing pair of linear programs. And in an optimal solution of the two problems, the optimal cost to the consumer, i.e., the optimal value z^*, *equals* the optimal return w^* to the supplier. This is the *strong duality* result.

Exercise: In the "shopping for food" problem of Section 1.1, assume that three different seafoods are available and that a single nutritional requirement on protein must be satisfied. Create hypothetical data (cost per pound and protein content per pound for each seafood), and then formulate the problem as a linear program. Find its optimal solution and justify your answer.

Computational Exercise: If you have access to software that solves linear programs, for example MATLAB and its optimization toolbox, use it to solve the foregoing two problems—"diet" and "price-setting"—and verify the weak and strong duality results. Show, in particular, that the optimal solution of the primal (diet) problem, to four-decimal accuracy, is

$$x_1^* = 2.5049, \quad x_2^* = 1.8385, \quad x_3^* = 4.6936, \qquad (5.1)$$

and that the optimal solution of the dual (price-setting) problem is

$$y_1^* = 1.6732, \quad y_2^* = 0.2140, \quad y_3^* = 0, \quad y_4^* = 4.6693, \quad y_5^* = 0. \qquad (5.2)$$

Verify that the optimal cost in the primal problem is $z^* = 174.7$, and this is the same, to working four-significant-figure accuracy, as the optimal income w^* in the dual problem.

There are many interesting observations that can be made about the foregoing optimal solutions. Thus for the price-setting problem, note that it is better to focus on getting as high a return as possible from three nutrients, protein, carbohydrate, and iron, even if this means having to give away the nutrients calcium and vitamin A without charge (so as not to violate any constraints of the associated linear program). One could obviously charge nonzero prices for all five nutrients and satisfy the constraints; for example, choose a small but nonzero number for each one. Then all constraints of the price-setting linear program will be satisfied, but the total return will be much less than the optimal profit w^* that can be obtained.

Note also that only the first, second, and fourth constraints of the primal, diet problem are satisfied as equalities at the optimal solution. The other two constraints, on calcium and vitamin A, are slack; i.e., one obtains more of these nutrients than the minimum daily requirement when the optimal purchase of foods is made. It is no coincidence that the corresponding variables of the dual, price-setting problem, the first, second, and fourth, are at nonzero level ("off their lower bounds"). This is called the *complementary slackness property* of the optimal solution. Recall that there are exactly as many dual variables as there are primal constraints, and let us give them corresponding indices; i.e., the first dual variable y_1 (price of protein) corresponds to the first primal constraint (on minimal daily protein requirement), and so on. In the optimal solution, observe that the only dual variables that are off their lower bounds, or "slack," are the ones that are associated with the constraints of the primal that hold as equalities (constraints that are "tight," or "not slack"). In like manner, the only variables of the primal that are off their lower bounds, or slack, in the optimal solution are the ones for which the corresponding dual constraints are tight, or not slack. In our example, all three primal variables have nonzero values and are thus slack, and the corresponding three dual constraints hold as equalities; i.e., they are not slack. Hence the term "complementary slackness." The justification for these observations will become clearer following the discussion of the next section.

5.2   Farkas Duality and LP Optimality

5.2.1   Weak Duality

Let us write the variables of the primal (diet) linear program and the coefficients of its linear objective function as row vectors $\mathbf{x}^T$ and $\mathbf{c}^T$, respectively, where

$$\mathbf{x}^T = [x_1, x_2, x_3], \quad \mathbf{c}^T = [40, 15, 10].$$

Let us also write the coefficients in the five inequality constraints of the primal problem as row vectors $\mathbf{a}_i^T$, $i = 1, \ldots, 5$, where

$$\mathbf{a}_1^T = [20, 3, 2], \quad \mathbf{a}_2^T = [0, 3, 18], \ldots, \mathbf{a}_5^T = [80, 7300, 0].$$

Then the primal can be restated as follows:

$$\begin{aligned}
\text{minimize } z = \quad & \mathbf{c}^T\mathbf{x} \\
\text{s.t.} \quad & \mathbf{a}_1^T\mathbf{x} \geq 65, \\
& \mathbf{a}_2^T\mathbf{x} \geq 90, \\
& \mathbf{a}_3^T\mathbf{x} \geq 200, \\
& \mathbf{a}_4^T\mathbf{x} \geq 10, \\
& \mathbf{a}_5^T\mathbf{x} \geq 5000,
\end{aligned}$$

$$x_1 \geq 0, \quad x_2 \geq 0, \quad x_3 \geq 0.$$

Since the columns of the dual problem correspond to the rows of the primal, the dual can be rewritten as follows:

$$\begin{aligned}
\text{maximize } w = \quad & 65y_1 + 90y_2 + 200y_3 + 10y_4 + 5000y_5 \\
\text{s.t.} \quad & \mathbf{a}_1 y_1 + \mathbf{a}_2 y_2 + \mathbf{a}_3 y_3 + \mathbf{a}_4 y_4 + \mathbf{a}_5 y_5 \leq \mathbf{c},
\end{aligned}$$

$$y_i \geq 0, \quad i = 1,\ldots,5.$$

Weak duality can be easily verified: (i) Multiply constraint i of the primal by y_i for each $i = 1,\ldots,5$ and sum all five constraint rows. (Note that $y_i \geq 0$ implies that the directions of the inequalities are unaltered.) (ii) Take the inner product of each of the columns of the dual with the vector $\mathbf{x}$. (Again, the components of $\mathbf{x}$ are nonnegative, so the directions of the inequalities are unaltered.) (iii) Compare the two expressions just obtained; note that they have the same left-hand sides; and follow the direction of the inequalities to immediately obtain the result $z \geq w$.

It will be convenient to define the row vector $\mathbf{y}^T$ of dual variables and the row vector $\mathbf{b}^T$ of right-hand sides of the primal as follows:

$$\mathbf{y}^T = [y_1, y_2, \ldots, y_5], \quad \mathbf{b}^T = [65, 90, 200, 10, 5000].$$

Then the weak duality result can be stated as

$$z \equiv \mathbf{c}^T\mathbf{x} \geq \mathbf{b}^T\mathbf{y} \equiv w.$$

The result tells us that the objective value associated with *any* feasible solution of the dual cannot exceed the objective value associated with *every* feasible solution of the primal. Similarly, the objective value associated with any feasible of the primal is at least as large as the objective value of every solution of the dual. Suppose we had a pair of feasible solutions in hand, one for the primal and the other for the dual, for which the associated objective function values were the same. Then weak duality informs us that these two feasible solutions must be *optimal* for their respective linear programs.

The foregoing observations do *not* rule out the possibility that the smallest (optimal) objective value over all feasible solutions of the primal is *strictly greater* than the largest (optimal) objective value over all feasible solutions of the dual. Hence the need for a strong duality result, which asserts that there is no *duality gap* between optimal solutions of the primal and dual linear programs. In other words, assuming that the primal and dual linear programs are not infeasible, i.e., their respective sets of inequality and bound constraints are not inconsistent, strong duality asserts that there must exist a feasible solution of the primal, say

$\mathbf{x}^*$, and a feasible solution of the dual, say $\mathbf{y}^*$, with associated objective values that are *equal*, i.e.,

$$z^* \equiv \mathbf{c}^T\mathbf{x}^* = \mathbf{b}^T\mathbf{y}^* \equiv w^*. \tag{5.3}$$

How do we find such a pair of feasible solutions? A classical lemma of Farkas [1902] makes the task easy—its significance for duality theory in optimization was recognized only several decades after its first appearance.

5.2.2 *Farkas' Lemma*

To introduce this key result, let us consider any given finite set of k vectors, $\mathbf{v}_1, \ldots, \mathbf{v}_k$, in n dimensions (denoted henceforth by R^n); for simplicity, take n to be 2 or 3, but of course n can be any positive integer. Suppose $\mathbf{u}$ is another given vector with the property that it can be expressed as a *nonnegative* linear combination of the vectors $\mathbf{v}_1, \ldots, \mathbf{v}_k$, i.e.,

$$\mathbf{u} = \alpha_1\mathbf{v}_1 + \alpha_2\mathbf{v}_2 + \cdots + \alpha_k\mathbf{v}_k, \quad \alpha_i \geq 0,\ i = 1, \ldots, k. \tag{5.4}$$

Take *any* vector $\mathbf{d}$ and form its inner product with the left- and right-hand side vectors in (5.4), i.e.,

$$\mathbf{u}^T\mathbf{d} = \sum_{i=1}^{k} \alpha_i(\mathbf{v}_k^T\mathbf{d}). \tag{5.5}$$

Since $\alpha_i \geq 0$, it follows immediately that

$$\mathbf{v}_i^T\mathbf{d} \geq 0,\ i = 1, \ldots, k,\ \text{implies that}\ \mathbf{u}^T\mathbf{d} \geq 0. \tag{5.6}$$

Farkas' lemma asserts that the *reverse statement* also holds: if the expression (5.6) is true for every vector $\mathbf{d}$, then the expression (5.4) must also be true. Let us state the result formally as follows:

Farkas' Lemma: Given a finite number of vectors $\mathbf{u}, \mathbf{v}_1, \ldots, \mathbf{v}_k$ in R^n, if the inequality $\mathbf{u}^T\mathbf{d} \geq 0$ is a consequence of the set of inequalities $\mathbf{v}_1^T\mathbf{d} \geq 0, \ldots, \mathbf{v}_k^T\mathbf{d} \geq 0$, then $\mathbf{u}$ is a nonnegative linear combination of $\mathbf{v}_1, \ldots, \mathbf{v}_k$.

We will not give a proof of this lemma here. Many proofs are available in the literature, and the simplest one that we have encountered is due to Komornik [1998]; see also the notes at the end of this chapter. The reader can obtain an intuitive feeling for the validity of the lemma from geometric pictures in two and three dimensions, which we recommend as an exercise.

5.2.3 *Strong Duality*

We can now turn to establishing the strong duality result. Given a primal feasible solution $\mathbf{x}^*$ that is claimed to be optimal, our task is to use this solution to identify a vector $\mathbf{y}^*$ that is feasible for the dual and that satisfies expression (5.3).

Observe first that not all the constraints of the primal are tight at $\mathbf{x}^*$. Only the first, second, and fourth primal constraints hold as equalities. Since the third and fifth constraints are slack at $\mathbf{x}^*$, any sufficiently small move away from $\mathbf{x}^*$ will give points that continue to satisfy these two constraints. Only the tight constraints have points that are infeasible in every neighbourhood of $\mathbf{x}^*$. The normals to these tight constraints, $\mathbf{a}_1, \mathbf{a}_2$, and $\mathbf{a}_4$, point in the directions of *increase* of the corresponding linear functions on the left-hand sides of the constraints.

Let us define a feasible direction $\mathbf{d}$ at the point $\mathbf{x}^*$ as one that points into the feasible region of the primal. Clearly any feasible direction must make a nonobtuse angle with normals of the tight constraints, i.e., feasible directions $\mathbf{d}$ must satisfy the three inequalities

$$\mathbf{a}_1^T \mathbf{d} \geq 0, \quad \mathbf{a}_2^T \mathbf{d} \geq 0, \quad \mathbf{a}_4^T \mathbf{d} \geq 0. \tag{5.7}$$

Since $\mathbf{x}^*$ is optimal, the primal objective function cannot decrease by taking a nonzero step along any feasible direction $\mathbf{d}$. The vector $\mathbf{c}$ points in the direction of greatest rate of increase of the objective function. Therefore, any feasible direction, i.e., any direction $\mathbf{d}$ satisfying (5.7), must also satisfy

$$\mathbf{c}^T \mathbf{d} \geq 0.$$

If we take $k = 3$ in Farkas' lemma, identify $\mathbf{u}$ with $\mathbf{c}$, and $\mathbf{v}_1, \ldots, \mathbf{v}_3$ with $\mathbf{a}_1, \mathbf{a}_2$, and $\mathbf{a}_4$, then we immediately obtain the result

$$\mathbf{c} = \alpha_1 \mathbf{a}_1 + \alpha_2 \mathbf{a}_2 + \alpha_4 \mathbf{a}_4, \quad \alpha_1 \geq 0, \ \alpha_2 \geq 0, \ \alpha_4 \geq 0. \tag{5.8}$$

Let us define the components of the vector $\mathbf{y}^*$ as follows:

$$y_1^* = \alpha_1, \quad y_2^* = \alpha_2, \quad y_3^* = 0, \quad y_4^* = \alpha_4, \quad y_5^* = 0. \tag{5.9}$$

The components of $\mathbf{y}^*$ obviously satisfy the inequality constraints of the dual, since the latter hold as equalities at the point $\mathbf{y}^*$; see (5.8). Also the components of $\mathbf{y}^*$ are nonnegative. Hence $\mathbf{y}^*$ is feasible for the dual. Its associated objective value is

$$w^* = 65y_1^* + 90y_2^* + 20y_3^* + 10y_4^* + 5000y_5^* = \mathbf{b}^T \mathbf{y}^*.$$

As just noted, the components y_i^* satisfy the inequality constraints of the dual as equations, and using (5.9), let us now write (5.8) in the form

$$\mathbf{c} = \mathbf{a}_1 y_1^* + \mathbf{a}_2 y_2^* + \mathbf{a}_3 y_3^* + \mathbf{a}_4 y_4^* + \mathbf{a}_5 y_5^*.$$

Take the inner product with $\mathbf{x}^*$ to obtain

$$\mathbf{c}^T \mathbf{x}^* = \sum_{i=1}^{5} y_i^* (\mathbf{a}_i^T \mathbf{x}^*). \tag{5.10}$$

Likewise, multiply each of the constraints of the primal problem by the corresponding nonnegative quantity y_i^* and use the fact that the first, second, and fourth constraints hold as equalities at the optimal point $\mathbf{x}^*$ and the third and fifth components of $\mathbf{y}^*$ are zero to obtain

$$\mathbf{b}^T \mathbf{y}^* = \sum_{i=1}^{5} y_i^* (\mathbf{a}_i^T \mathbf{x}^*). \tag{5.11}$$

Then combining (5.10) and (5.11), we immediately obtain the strong duality result

$$z^* \equiv \mathbf{c}^T \mathbf{x}^* = \mathbf{b}^T \mathbf{y}^* \equiv w^*.$$

Note that we have needed symbolic values only for the components of $\mathbf{y}^*$, which are derived from Farkas' lemma. But if their numerical values are sought, they must obviously be given by (5.2). Also, we have obtained the strong duality result for a particular pair of dual linear programs, but the proof for an arbitrary linear program and its dual requires very little in the way of modification.

We see also that duality results and optimality conditions for a linear program are different sides of the *same* coin. For example, we could state optimality conditions for the primal, diet problem as follows: a given vector $\mathbf{x}^*$ is optimal for the diet problem if and only if

(i) $\mathbf{x}^*$ is primal feasible;

and there exists a vector $\mathbf{y}^*$ such that

(ii) $\mathbf{y}^*$ is dual feasible;

(iii) $\mathbf{c}^T \mathbf{x}^* = \mathbf{b}^T \mathbf{y}^*$.

A variant is to replace (iii) by so-called complementary slackness conditions outlined at the end of Section 5.1. For further details on this equivalence, see, for example, Chvátal [1983].

A general linear program has a linear objective that may be maximized or minimized, and its constraints may be a mix of equalities, $\geq$ inequalities, and $\leq$ inequalities. Instead of simple nonnegativity bounds, its variables may be required to satisfy more general lower and upper bounds, for example, $1 \leq x_1 \leq 10$. Every such linear program has a corresponding dual, and weak and strong duality results will hold for the pair. These observations can easily be validated by defining a *canonical* or *standard* form for the primal, for example, the one to which the diet linear program conforms, and then converting any given linear program to the chosen canonical form by means of simple transformations; see also Section 6.1.2.

5.3   Notes

Sec 5.2: The relevance of Farkas' lemma (Farkas [1902]) for optimization was noted, at first, in the classic work of Von Neumann and Morgenstern [1944] on game theory. Following conversations with John Von Neumann, duality results for linear programming were obtained informally by G.B. Dantzig (see his historical account in the foreword of Dantzig and Thapa [1997]), and formally by Gale, Kuhn, and Tucker (see their article in the book edited by Koopmans [1951]).

6

The Golden Age of Optimization

The genealogy of optimization can be traced to the works of venerated mathematicians: Cauchy, Euler, Fourier, Gauss, Kantorovich, Lagrange, Newton, Poincaré, Von Neumann, and others. But it was only after a very long gestation period that the subject was truly born in the mid-nineteen forties with G.B. Dantzig's discovery of the wide-ranging practical applicability of the linear programming model—the flagship of the field—and his invention of its main solution engine, the *simplex algorithm*. Other areas of mathematical programming developed rapidly in what may fittingly be termed the *Dantzig Modeling-and-Algorithmic Revolution*. Its history is nicely told in the foreword of the book by Dantzig and Thapa [1997]; see also the optimization classic, Dantzig [1963].

In the previous chapter, we described the linear programming model and its underlying duality theory. Now we introduce the reader to the simplex algorithm and discuss its use in practice. We also outline its important specialization for solving network flow problems of Chapter 4 via their linear programming equivalents.

6.1 Dantzig's Simplex Algorithm

6.1.1 Example in 3-D

Consider a very simple instance of a linear program in three-dimensional space (henceforth R^3) as follows: minimize a linear objective function $x_1 + x_2 + x_3$ where the variables x_1, x_2, and x_3 must satisfy the single linear equality constraint $3x_1 + 6x_2 + 4x_3 = 12$ and are constrained to be nonnegative. The points that are feasible for this linear program must lie within the intersection of the plane defined by the equality constraint and the nonnegative constraints $x_j \geq 0$, $j = 1, 2, 3$. This linear program can be written more compactly as follows:

$$\text{minimize } \mathbf{c}^T \mathbf{x}$$

$$\text{s.t. } \mathbf{a}^T \mathbf{x} = b, \tag{6.1}$$

$$x_1 \geq 0, \quad x_2 \geq 0, \quad x_3 \geq 0,$$

where $\mathbf{c}^T = [1, 1, 1]$ and $\mathbf{a}^T = [3, 6, 4]$ are row vectors; $\mathbf{x}$ is a column vector with components x_1, x_2, and x_3; and $b = 12$.

For purposes of discussion, it is convenient to be a little more general. Thus, let $\mathbf{c}^T = [c_1, c_2, c_3]$, where the components $c_j, j = 1, 2, 3$, are three given real numbers (with unrestricted signs). Similarly, let $\mathbf{a}^T = [\beta, \eta_1, \eta_2]$, where the components are three given *positive* real numbers.[1] The righthand side b is a given positive number; and the three variables $x_j, j = 1, 2, 3$, are constrained to be nonnegative. Figure 6.1 depicts the triangular feasible region: the set of points that satisfy the constraints of the linear program.

Exercise: Draw the version of Figure 6.1 corresponding to the specific choices for **c**, **a**, and b given immediately after the linear program (6.1). Also draw contours of the objective function within the feasible region, i.e., lines on which $\mathbf{c}^T \mathbf{x}$ has a constant value, and use them to locate the optimal solution of the linear program.

The simplex algorithm, when applied to the LP example of Figure 6.1, proceeds as follows: Start at any vertex of the triangular feasible region, say $\mathbf{x}^{(0)}$, find an edge along which the objective function decreases, and move along it to an adjacent vertex. Then repeat until no further progress can be made, which can happen only at the optimal solution. The vertex $\mathbf{x}^{(0)}$ and vectors $\mathbf{z}_1$ and $\mathbf{z}_2$ that lie along the edges leading to the two adjacent vertices are as follows:

$$\mathbf{x}^{(0)} = \begin{bmatrix} \beta^{-1} b \\ 0 \\ 0 \end{bmatrix}, \quad \mathbf{z}_1 = \begin{bmatrix} -\beta^{-1} \eta_1 \\ 1 \\ 0 \end{bmatrix}, \quad \mathbf{z}_2 = \begin{bmatrix} -\beta^{-1} \eta_2 \\ 0 \\ 1 \end{bmatrix}, \tag{6.2}$$

where $\beta^{-1} b > 0$, and the horizontal and vertical axes in Figure 6.1 correspond to the variables x_1 and x_2, respectively.

The scalar products between the vector **c** and the vectors $\mathbf{z}_1$ and $\mathbf{z}_2$ define the *rates of change* of the linear objective function along the edges:

---

[1]We employ a notation for the components that is convenient for subsequent exposition; see Section 6.1.2. Note also that we use boldface letters to denote vectors and italic letters for their components and other scalar quantities.

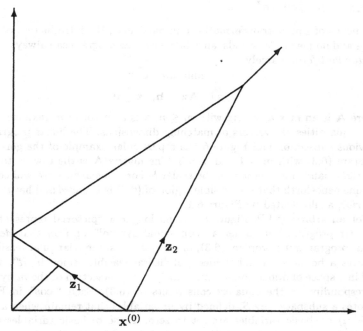

FIGURE 6.1. LP example

positive if the objective is increasing and negative if decreasing. (Formally, these quantities are the *directional derivatives* along the edges.) One can proceed along any "improving" edge whose corresponding rate of change is negative. For example, one could choose the *steepest edge* as follows: Normalize each vector z_i, $i = 1, 2$, by dividing it by its length $\|z_i\|$, compute the directional derivatives along the normalized vectors, and choose the most negative. In Figure 6.1, this would be, say, the edge z_1. Note also that the vectors z_1 and z_2 are orthogonal to a. Proceed along an improving edge to a new vertex. (In order to obtain a situation at the new, or current, vertex that is identical to the one just described, simply reorder the variables and denote the current vertex by $x^{(0)}$.) Then repeat the procedure. This is the essence of the simplex algorithm. It is terminated when there are *no* improving edges at the current vertex, which is the optimal solution.[2] The latter will be found after at most two iterations on this simple example.

Exercise: Apply the simplex algorithm to the specific linear program (6.1), starting at the vertex with components $(4, 0, 0)$. Find another choice of vector c so that the algorithm takes two iterations to terminate.

---

[2]The simplex algorithm demonstrates constructively that *an* optimal solution of the linear program occurs at a vertex of the feasible region.

6.1.2 In General

By means of simple transformations, in particular, the introduction of new variables and nonnegative bounds, an arbitrary linear program can always be restated in *standard form*, namely,

$$\text{minimize } \mathbf{c}^T \mathbf{x}$$

$$\text{s.t. } \mathbf{Ax} = \mathbf{b}, \ \mathbf{x} \geq \mathbf{0}, \tag{6.3}$$

where $\mathbf{A}$ is an $m \times n$ matrix with $m \leq n$, $\mathbf{x}$ is a vector of n unknowns, and the other quantities are vectors of matching dimensions. The linear program of the previous subsection and Figure 6.1 is a particular example of the general linear program (6.3) with $m = 1$ and $n = 3$. The matrix $\mathbf{A}$ is the row vector $\mathbf{a}^T$, and the right-hand-side vector $\mathbf{b}$ is the scalar b. For convenience, we will continue to assume henceforth that the feasible region of (6.3) is *bounded* and has a *nonempty interior*, as illustrated by Figure 6.1.

For an arbitrary LP, Figure 6.1 is no longer a "pictorial representation" of a linear program but rather a "compound symbol" or *icon* that *denotes* the linear programming problem (6.3), in general Thus, the triangular feasible region denotes a bounded or unbounded polyhedral feasible region in R^n, which lies within a space of dimension $n - m$, namely, the intersection of the m hyperplanes corresponding to the equality constraints of (6.3). Each "axis" in Figure 6.1 denotes a subspace, say S, defined by m variables that remain when a subset of $(n - m)$ nonbasic variables are set to zero. These m basic variables must also satisfy the m equality constraints, and their values are therefore determined. When these values are nonnegative, then the basic and nonbasic variables define a vertex of the feasible region that lies in the subspace S. Note that there is an *enormous* number of such "axes," an upper bound U being the number of ways of choosing m objects from a set of n objects, namely,

$$U = \frac{n(n-1)(n-2)\cdots(n-m+1)}{m(m-1)(m-2)\cdots 1}.$$

For even moderate choices of the dimensions n and m, the number U becomes exponentially large. The following exercises will clarify the foregoing discussion:

Exercise: Consider the two inequality constraints

$$x_1 + x_2 \leq 10,$$

$$-2x_1 + 3x_2 \geq 6.$$

Verify that they can be converted into equalities by adding a slack variable $s_1 \geq 0$ to the first constraint and subtracting a slack variable $s_2 \geq 0$ from the second constraint.

Exercise: Add a second linear *equality* constraint to the simple example of Section 6.1.1 and revise Figure 6.1 to give a picture of this new linear program.

Exercise: Compute the upper bound U for the choice of dimensions $n = 50$ and $m = 20$.

Exercise: In analogy to Figure 6.1, the feasible region of the general linear program, which is formally called a *convex polyhedral set*, is defined by "billions and billions[3] of axes," each a subspace of dimension m that may contain a vertex. Try to envision a geometric object of such extraordinary complexity, for which Figure 6.1 serves now *only as an icon*.

Notwithstanding our inability to visualize the feasible region geometrically, except in two or three dimensions, the description of the simplex algorithm, applied to the general linear program (6.3), remains *in essence* no different from its description on the very simple example depicted in Figure 6.1. Partition the columns of the LP matrix A into $A = [B, N]$, where B is an $m \times m$ *basis matrix* assumed to be nonsingular and defined by the first m columns of A, and N is an $m \times (n - m)$ matrix consisting of the remaining columns of A. (In the earlier example, $m = 1$, $n = 3$, the matrix B corresponds to the 1×1 matrix β, and N to the 1×2 matrix $[\eta_1, \eta_2]$.) The starting vertex is given by a vector $x^{(0)}$, and the edges leading from it are defined by the $n - m$ columns of a matrix Z as follows:

$$x^{(0)} = \begin{bmatrix} B^{-1}b \\ 0 \end{bmatrix}, \quad Z = \begin{bmatrix} -B^{-1}N \\ I \end{bmatrix}, \quad (6.4)$$

where 0 denotes a zero vector of appropriate dimension, and I denotes the $(n-m)$ by $(n - m)$ identity matrix. We assume that the basis matrix B is such that $B^{-1}b \geq 0$. The *directional derivatives* along edges are given by the elements of the vector $c^T Z$. Again one can proceed to an improving vertex along any edge with a negative directional derivative. The operations of the simplex method now involve inversion of the basis matrix, which is usually handled implicitly by forming and updating matrix factorizations and solving associated systems of linear equations.

At a vertex, the set of m equality constraints in (6.3) and the set of $(n - m)$ nonnegative bound constraints that hold as equalities (the corresponding nonbasic variables are fixed at value zero) then jointly comprise a set of n constraints that are said to be *active* at the point $x^{(0)}$. The directional derivatives along edges, namely $c^T Z$, correspond to the $(n - m)$ active bound constraints at $x^{(0)}$. In the simplex algorithm, one of these active bound constraints, whose directional derivative, or reduced cost as it is often called, has a negative sign, is relaxed. A move is then made along the corresponding edge to a new vertex, which is again defined by a set of n active constraints.

Under appropriate so-called *nondegeneracy* assumptions, upon which we will not elaborate here, the simplex algorithm can easily be shown to converge to the optimal solution of a linear program in a *finite* number of iterations. If such assumptions are violated, then a very elegant rule[4] proposed by Bland [1977] is the simplest among several techniques for selecting entering and exiting variables

---

[3]A phrase that the late Carl Sagan made his own—through repeated use in his famous television series on astronomy—when describing, in wonderment, the number of stars in the firmament.

[4]Assume a fixed ordering of the variables and resolve ties in the entering or exiting variables by always choosing the candidate with the *smallest* index in the ordering.

that guarantee convergence of the algorithm, in general. For details, see, for example, Chvátal [1983].

In order to implement the simplex algorithm, observe the following:

- In expression (6.4), obtaining the *nonzero components* of the vertex $\mathbf{x}^{(0)}$, which we denote by $\mathbf{x}_B^{(0)}$, requires the solution of a linear system of equations

$$\mathbf{Bx}_B^{(0)} = \mathbf{b}. \tag{6.5}$$

- The rates of change along edges, i.e., the directional derivatives $\mathbf{c}^T\mathbf{Z}$, can be written as follows:

$$\mathbf{c}^T\mathbf{Z} = [\mathbf{c}_B^T, \mathbf{c}_N^T] \begin{bmatrix} -\mathbf{B}^{-1}\mathbf{N} \\ \mathbf{I} \end{bmatrix} = [\mathbf{c}_N^T - (\mathbf{c}_B^T\mathbf{B}^{-1})\mathbf{N}],$$

where $\mathbf{c}_B^T$ and $\mathbf{c}_N^T$ denote the first m and remaining $n - m$ components of $\mathbf{c}^T$, respectively. The quantity $\mathbf{w}^T = \mathbf{c}_B^T\mathbf{B}^{-1}$ can then be obtained by solving the system of linear equations

$$\mathbf{w}^T\mathbf{B} = \mathbf{c}_B^T \text{ or equivalently } \mathbf{B}^T\mathbf{w} = \mathbf{c}_B. \tag{6.6}$$

Note that it would be inefficient to first compute all edge vectors and then compute their associated directional derivatives. All that is needed is the single vector $\mathbf{w}$, followed by computation of $(\mathbf{c}_N^T - \mathbf{w}^T\mathbf{N})$.

- After the edge along which improvement can be made is identified implicitly using the directional derivatives just computed, the edge itself must be found. This is one of the columns of $\mathbf{Z}$, and the corresponding column in $\mathbf{N}$ is, say, $\mathbf{a}_k$. The edge vector is easily obtained from $\mathbf{y} = \mathbf{B}^{-1}\mathbf{a}_k$, which requires the solution of a third system of linear equations

$$\mathbf{By} = \mathbf{a}_k. \tag{6.7}$$

Thus when viewed purely in *implementational* terms, the simplex algorithm requires *the coordinated solution of three systems of linear equations*, namely, (6.5)–(6.7). Each involves the same basis matrix $\mathbf{B}$. These three linear systems are changed at each iteration, by revising the basis matrix, one column at a time. Artful representation of the basis matrix in a factored form and the updating of this factorization at each iteration can lead to enormous gains in efficiency within the algorithm when solving the foregoing linear systems.

6.2 Linear Programming in Practice

It is when one turns to *practical* linear programs and formulates the foregoing procedures into effective *numerical* algorithms that linear programming comes into its own. In a practical linear program (6.3), the matrix $\mathbf{A}$ could have thousands of rows and tens of thousands of columns and would generally be very sparse; i.e., a typical column would contain only a few nonzero elements, often between 5 and 10. (In a practical linear program, the non-negativity bounds are also often replaced by upper and lower bounds on the variables.) Two considerations are then of paramount importance:

- How many iterations, as a function of problem dimensions, do our methods take on typical problems?

- How efficient can each iteration be made?

These *complexity* considerations are what distinguish the *numerical* algorithm from the *mathematical* algorithm of the previous sections.

In the case of the simplex method, the *mathematical* algorithm, described in terms of vertex following, was proposed independently at least three different times. The contributions of G.B. Dantzig to the *numerical* algorithm are best described in his own words; see Dantzig [1983]:

> It is my opinion that any well trained mathematician viewing the linear programming problem in the row geometry of the variables would have immediately come up with the idea of solving it by a vertex descending algorithm as did Fourier, de la Valée Poussin, and Hitchcock before me—each of us proposing it independently of the other. I believe, however, that if anyone had to consider it as a practical method, as I had to, he would quickly have rejected it on intuitive grounds as a very stupid idea without merit. My own contributions towards the discovery of the simplex method were (1) independently proposing the algorithm, (2) initiating the software necessary for its practical use, and (3) observing by viewing the problem in the geometry of the columns rather than the rows that, contrary to geometric intuition, following a path on the outside of the convex polyhedron might be a very efficient procedure.

Item (3) above represents one of the early contributions to *complexity analysis*. In practice, the typical number of iterations of the simplex algorithm increases proportionally to m, with a very small constant of proportionality (for example, between 1.5 and 3). The arguments that led Dantzig to this fact were later formalized by him; see, for example, Dantzig [1980]. The "LP technology" initiated by Dantzig relies on and, indeed, stimulated the development of appropriate techniques for solving large and very sparse systems of linear equations. These techniques are based on factorization of the matrix, and they implicitly perform the basis matrix inversion involved in (6.4); see also (6.5)–(6.7). (They are called direct methods because they operate directly on the basis matrix $\mathbf{B}$ in order to obtain a representation of it that permits very efficient solution of the associated systems of linear equations.) These then were the two key contributions of Dantzig to the simplex method, in addition to proposing the mathematical algorithm independently of earlier historical antecedents, as mentioned in the above quotation. *It is important to appreciate the quantum leap from a viable mathematical algorithm to a viable numerical algorithm*, which is why Dantzig's name is synonymous with the simplex method. Subsequent to the acceptance of the simplex method as the workhorse of linear programming,

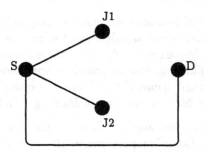

FIGURE 6.2. Spanning tree defining a basis matrix

there was an explosion of development that led to numerous variants, for example, the so-called dual, primal-dual, and self-dual simplex algorithms. It is also worth noting that linear programming as we know it today would not have been possible without the digital computer, and developments in the two fields advanced together in tandem; see Dantzig [1985].

6.3   Network Simplex Algorithm

In earlier chapters, we have seen examples of linear programs that arise from networks, and in particular, that their LP matrices are very sparse, having just two nonzero numbers in each column. The basis matrices of these linear programs have remarkable properties that greatly facilitate the solution of associated linear systems of equations in the main operations of the simplex algorithm. This is the "bottom-up" or continuous approach to developing algorithms for solving the network flow problems of Section 4.3.

Consider specifically an optimal cost/profit network flow problem. We can note the following:

- A basis matrix of the equivalent linear program corresponds to the arcs of a spanning tree of the network (with directions on arcs ignored); see Section 4.2 for the definition of a spanning tree. For example, consider the maximum profit network flow problem of Section 4.1.1 and Figure 4.1, and its equivalent linear program described in Section 4.1.2. An example of a spanning tree for the problem of Figure 4.1 (with the directions on arcs of the network removed to obtain the associated undirected network) is shown in Figure 6.2. Each arc of a spanning tree defines a column of an associated basis matrix.

- The columns chosen in this way can be shown to be linearly independent. An additional artificial column, with a unit element in the last position and zeros elsewhere, is added as the last column to obtain a nonsingular basis matrix. This corresponds to adding an artificial "root" arc to the network flow problem of Figure 4.1 and the spanning tree of Figure 6.2.

- It can also be shown that is it always possible to reorder, or permute, the rows and columns of the basis matrix into a *triangular* matrix. A linear system of equations whose matrix of coefficients is in triangular form is very easy to solve by forward (or back) substitution. (These observations can be verified easily for the example in Figure 6.2, and this is left as an exercise for the reader.)

- The basis matrix and associated linear systems do *not* need to be formed or permuted explicitly. Their forward or back substitution operations can be performed *on the network itself*, as can all the other key operations of the simplex algorithm.

The foregoing facts result in enormous enhancements in efficiency. This is the so-called *technology of the simplex algorithm*: its adaptation to a variety of network flow problems via their linear programming equivalents. For a comprehensive discussion, see, for instance, Bazaraa et al. [1990]; and for a brief and very readable account, see, in particular, Evans and Minieka [1992].

6.4   Notes

Sec. 6.0: An earlier version of the foreword of Dantzig and Thapa [1997] can be found in Dantzig's article in Lenstra et al. [1991]. See also the *Encyclopedia of Optimization*, edited by Floudas and Pardalos [2001], for a wide-ranging overview of the entire field of optimization.

7

An Algorithmic Revolution

In 1984, coincidentally the Orwellian[1] year, linear programming was considered to be a well-settled subject: combinatorially oriented, closely allied to network flow programming, and capable of being pursued in a largely self-contained manner. The remarkable practical efficiency of its main solution engine, the simplex algorithm, which traverses a sequence of vertices across the boundary of the feasible region and is thus sometimes termed an "exterior" method, had at last been given a satisfactory theoretical explanation (in the late 1970s). Around the same time, another important theoretical issue was put to rest. The LP problem was shown to be solvable in "polynomial time" by using an alternative "ellipsoidal" algorithm (Khachiyan [1979]). This demonstrated the existence of an algorithm that was superior to the simplex algorithm in theory (in the sense of "worst-case," as contrasted with "average-case," behaviour). But in practice, the ellipsoidal algorithm proved to be inefficient, and not at all competitive with the simplex algorithm.

The situation changed dramatically following the publication, in 1984, of a seminal paper on algorithmic linear programming by N. Karmarkar. This contribution pointed the way to new "interior-point" linear programming algorithms that closed the theory–practice gap mentioned above: they were both polynomial-time in theory and efficient in practice. During the subsequent decade, the subject of linear programming entered a fresh and exciting cycle of growth and development. The membrane that had long

---

[1] *1984* is the title of George Orwell's futuristic novel.

separated linear programming from the enveloping subject of nonlinear programming was largely breached. New life was breathed into existing optimization techniques, some of which had fallen into disuse. And the repertoire of algorithms for solving practical linear programs was significantly broadened. But in an interesting turn of events, "simplex technology" also received a significant boost during this post-Karmarkar period, and the primacy of the simplex algorithm, as the practical LP solution technique of first choice, was *not* supplanted.

Among the many new discoveries, two proved to be of paramount importance to the ensuing algorithmic revolution: *affine scaling* and the *central path* of a linear program. In this chapter, we provide an introduction to these two foundational ideas and to recent interior-point algorithms for linear programming that are premised on them.

7.1   Affine Scaling

7.1.1   Example in 3-D

Let us return to the example of Section 6.1.1. The line shown within the triangular region in Figure 7.1 is a "feasible contour" of the linear objective function, i.e., a set of points within the feasible region for which the value of the objective function $\mathbf{c}^T\mathbf{x}$ is the same. Other contour lines[2] are parallel to the one depicted and have different associated objective function values.

Suppose that an approximation to the solution, or *current iterate*, is at an *interior point*,

$$\mathbf{x}^{(1)} = \left(x_1^{(1)}, x_2^{(1)}, x_3^{(1)}\right), \quad x_j^{(1)} > 0, \ j = 1, 2, 3,$$

i.e., a point that lies within and strictly off the boundary of the feasible region. The value of the objective function at the current iterate, and at all points on the depicted contour line through it, is $z^{(1)} = \mathbf{c}^T\mathbf{x}^{(1)}$.

In Figure 7.1, the vector $\mathbf{d}^{(1)}$, in the plane $\mathbf{a}^T\mathbf{x} = b$ and at right angles to the contour line, points in the feasible direction of the most rapid rate of decrease of the objective function at the iterate $\mathbf{x}^{(1)}$. (In the next subsection, we describe how to obtain this vector algebraically, and at present, we simply assume it is obtained by graphical means.) One can compute the step length along this direction, from $\mathbf{x}^{(1)}$ to the boundary, which is given by the largest value of α for which all components of $\mathbf{x}^{(1)} + \alpha\mathbf{d}^{(1)}$ remain nonnegative; call this quantity α_{max}. Choose a step length along the direction that is slightly smaller than α_{max} in order to obtain a new iterate, say $\mathbf{x}^{(2)}$, that again lies in the interior of the feasible triangle, i.e.,

---

[2]Note that contour lines are always orthogonal to the vector $\mathbf{c}$, and the latter need not lie within the plane defined by $\mathbf{a}^T\mathbf{x} = b$.

the components of $\mathbf{x}^{(2)}$ must remain strictly positive. For the particular choice of $\mathbf{x}^{(1)}$ shown in Figure 7.1, good progress will be made during the first iteration. But there is no point in repeating the procedure at $\mathbf{x}^{(2)}$, because the same direction of "steepest feasible descent" would be obtained again, and subsequent iterates would jam near the boundary.

The affine-scaling algorithm is based on a novel yet incredibly simple modification of this approach; terrific ideas are often very simple in hindsight! This is as follows: *change the metric or measure of distance by a rescaling of variables.* Let $\mathbf{x}^{(k)} = \left(x_1^{(k)}, x_2^{(k)}, x_3^{(k)} \right)$ denote a current iterate,[3] where $k \geq 1$. Let us rescale the variables using the components of this current iterate and define new variables

$$\tilde{\mathbf{x}} = (\tilde{x}_1, \tilde{x}_2, \tilde{x}_3) = \left(\frac{x_1}{x_1^{(k)}}, \frac{x_2}{x_2^{(k)}}, \frac{x_3}{x_3^{(k)}} \right). \tag{7.1}$$

Note that $\tilde{\mathbf{x}}^{(k)} = (1, 1, 1)$. In the transformed "tilde" space, the linear program (6.1) becomes

$$\text{minimize } \tilde{\mathbf{c}}^T \tilde{\mathbf{x}}$$

$$\text{s.t. } \tilde{\mathbf{a}}^T \tilde{\mathbf{x}} = b, \ \tilde{\mathbf{x}} \geq \mathbf{0},$$

where

$$\tilde{\mathbf{c}}^T = \left[c_1 x_1^{(k)}, c_2 x_2^{(k)}, c_3 x_3^{(k)} \right] \text{ and } \tilde{\mathbf{a}}^T = \left[\beta x_1^{(k)}, \eta_1 x_2^{(k)}, \eta_2 x_3^{(k)} \right],$$

using the notation of Section 6.1.1 for the original vectors $\mathbf{c}$ and $\mathbf{a}$.

Exercise: For the specific example at the beginning of Section 6.1.1 and an interior current iterate $\mathbf{x}^{(k)}$ of your choice, determine the *"scaled"* linear program in the "tilde" space, where the transformed current iterate is $(1, 1, 1)$. Depict its feasible region graphically, analogously to Figure 7.1.

The direction of most rapid feasible descent, say $\tilde{\mathbf{d}}^{(k)}$, is again orthogonal to the feasible contour lines in the transformed space. In the original space, the corresponding direction, $\mathbf{d}^{(k)}$, is obtained from a restoration of variables, or the inverse of the transformation (7.1), and its components are $d_j^{(k)} = x_j^{(k)} \tilde{d}_j^{(k)}$, $j = 1, 2, 3$. Reverting to the original variables in Figure 7.1, this direction is depicted at the iterate $\mathbf{x}^{(2)}$. We see that the net effect of affine scaling is to bend the search direction $\mathbf{d}^{(2)}$ away from the boundary of the feasible region, so that greater progress along the direction is possible. Take an improving step, subject again to remaining in the interior of the feasible region. Then repeat the procedure. This, in essence, defines

---

[3]For convenience, we now use "tuple" notation, specifically a 3-tuple, to identify the components of $\mathbf{x}^{(k)}$. Alternatively, $\mathbf{x}^{(k)T} = \left[x_1^{(k)}, x_2^{(k)}, x_3^{(k)} \right]$.

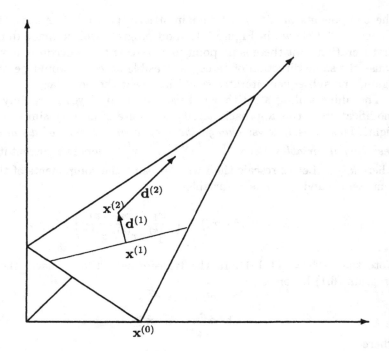

FIGURE 7.1. Affine scaling

the primal affine-scaling algorithm. It is terminated when progress can no longer be made in reducing the objective function, or changing the current iterate, as defined by a small convergence tolerance.

7.1.2   In General

Let us return to the initial iterate $\mathbf{x}^{(1)} > \mathbf{0}$ of Figure 7.1. The direction of steepest feasible descent $\mathbf{d}^{(1)}$ at this iterate was obtained graphically in the previous subsection. It can instead be defined algebraically as follows: let

$$\mathbf{P} = \mathbf{I} - \mathbf{a}(\mathbf{a}^T\mathbf{a})^{-1}\mathbf{a}^T = \mathbf{I} - \frac{\mathbf{a}\mathbf{a}^T}{\mathbf{a}^T\mathbf{a}} \qquad (7.2)$$

be the orthogonal projection matrix $\mathbf{P}$ into the plane $\mathbf{a}^T\mathbf{x} = b$. Then

$$\mathbf{d}^{(1)} = \mathbf{P}(-\mathbf{c}) = -\mathbf{c} + \frac{\mathbf{a}^T\mathbf{c}}{\mathbf{a}^T\mathbf{a}}\mathbf{a}.$$

For the general linear program in standard form (6.3), the orthogonal projection matrix $\mathbf{P}$ into the "hyperplane" defined by the m equality constraints $\mathbf{A}\mathbf{x} = \mathbf{b}$ is defined analogously by

$$\mathbf{P} = \mathbf{I} - \mathbf{A}^T(\mathbf{A}\mathbf{A}^T)^{-1}\mathbf{A}. \qquad (7.3)$$

The equivalence between (7.3) and the corresponding quantity (7.2) is obtained by making the identification $\mathbf{A} \equiv \mathbf{a}^T$. Then $\mathbf{d}^{(1)} = \mathbf{P}(-\mathbf{c})$, the projected negative-gradient direction, which is orthogonal to the feasible contours, is a natural choice for the improving direction.

The affine-scaling algorithm as described in the previous subsection is essentially unchanged in the more general setting, but it requires a little more algebraic machinery. Suppose $\mathbf{x}^{(k)} > \mathbf{0}$, $k \geq 1$, denotes the current, interior-point iterate, and let $\mathbf{X}_k$ denote the *diagonal* matrix whose diagonal elements are the components of $\mathbf{x}^{(k)}$, i.e., $\mathbf{X}_k = \text{diag}[x_1^{(k)}, \dots, x_n^{(k)}]$. Make the diagonal transformation, or scaling, of variables $\mathbf{x} = \mathbf{X}_k \tilde{\mathbf{x}}$. Thus the transformed linear program is as follows: minimize $\tilde{\mathbf{c}}^T \tilde{\mathbf{x}}$ subject to $\tilde{\mathbf{A}} \tilde{\mathbf{x}} = \mathbf{b}$, $\tilde{\mathbf{x}} \geq \mathbf{0}$, where $\tilde{\mathbf{c}} = \mathbf{X}_k \mathbf{c}$ and $\tilde{\mathbf{A}} = \mathbf{A} \mathbf{X}_k$. Compute the projected direction $\tilde{\mathbf{d}}^{(k)}$ as we did above, but now in the transformed space, i.e., $\tilde{\mathbf{d}}^{(k)} = \tilde{\mathbf{P}}(-\tilde{\mathbf{c}})$, where $\tilde{\mathbf{P}}$ is given by

$$\tilde{\mathbf{P}} = \mathbf{I} - \tilde{\mathbf{A}}^T (\tilde{\mathbf{A}} \tilde{\mathbf{A}}^T)^{-1} \tilde{\mathbf{A}}.$$

In the original space this corresponds to the direction $\mathbf{d}^{(k)} = \mathbf{X}_k \tilde{\mathbf{d}}^{(k)} = -\mathbf{X}_k \tilde{\mathbf{P}} \mathbf{X}_k \mathbf{c}$, and in terms of the quantities that define the original linear program we have

$$\mathbf{d}^{(k)} = -\mathbf{X}_k [\mathbf{I} - \mathbf{X}_k \mathbf{A}^T (\mathbf{A} \mathbf{X}_k^2 \mathbf{A}^T)^{-1} \mathbf{A} \mathbf{X}_k] \mathbf{X}_k \mathbf{c}, \ k \geq 1. \tag{7.4}$$

The quantity $(\mathbf{A} \mathbf{X}_k^2 \mathbf{A}^T)^{-1}$ is an $m \times m$ matrix, and it represents the most taxing part of the computation. Implicitly or explicitly, the operations of the affine-scaling algorithm involve a matrix inversion, as in the simplex algorithm.

As noted above, the net effect of affine scaling is to bend the search direction away from the boundary of the feasible region so that greater progress is possible. Take an improving step subject to remaining in the interior of the feasible region. Then repeat the procedure. This defines the primal affine-scaling algorithm for the linear program (6.3). Under appropriate nondegeneracy assumptions,[4] the sequence of iterates can be shown to converge, i.e., $\mathbf{x}^{(k)} \to \mathbf{x}^*$.

Extensive numerical studies have revealed that the affine-scaling algorithm usually takes a remarkably small number of iterations to reach a solution: an oft-quoted, rule-of-thumb estimate is the logarithm of n, for any realistic value of the dimension n. The cost of each iteration of the algorithm is dominated by the cost of computing the search direction via an implicit matrix inversion in expression (7.4). Earlier, we noted that the simplex algorithm, which traverses a path of vertices on the boundary of the feasible region, also converges, in general, in a remarkably small number of iterations: a rule-of-thumb estimate was given in Section 6.2. Furthermore, an iteration of the simplex algorithm is significantly cheaper than an iteration of the affine-scaling algorithm. This is why the simplex algorithm is so hard to beat in practice.

Exercise: Consider a linear program in standard form (6.3) with dimensions m and $n = 2m$. Suppose an estimate of the number of iterations required to solve this linear program using the affine-scaling algorithm is $k_1 \ln(n)$, where $\ln$ denotes the

---

[4]The proof is technically difficult. It is also possible to show convergence in the absence of nondegeneracy assumptions, but then the proof is even more challenging to obtain.

natural logarithm and k_1 is a small integer. Suppose an estimate of the number of iterations to solve to solve the linear program by the simplex algorithm is $k_2 m$, where k_2 is another small integer. Explore the *costs per iteration* of the simplex and affine-scaling algorithms for solving the foregoing linear program, and then consider what the cost per iteration of the simplex algorithm must be *relative* to the cost per iteration of the affine-scaling algorithm in order that the former algorithm be more efficient than the latter.

7.2   Central Path

7.2.1   Example in 3-D

As we have noted, the simplex algorithm traverses points on the boundary of the feasible region. The affine-scaling algorithm remains in the interior, but its iterates can quickly approach the boundary. Affine scaling then provides a mechanism for turning a search direction away from the boundary, as shown in the example of Figure 7.1, but the sequence of iterates generated will often continue to lie close to it. A way to steer iterates back into the interior is needed. This is provided by the central path of a linear program: a *guiding thread* within the interior of the feasible region.

To introduce the central path, let us again consider the example of Section 6.1.1 in 3 dimensions. We seek a function that becomes smaller as points where it is evaluated approach the boundary of the feasible region, and that becomes larger as points move further away from the boundary. A simple function of this type, which comes almost immediately to mind, is of the form $x_1 x_2 x_3$; we will see others later.[5] This function has value zero on the boundary where $x_i = 0$ for at least one variable; it is positive on the interior of the feasible region; and it increases with each variable. When the function $x_1 x_2 x_3$ is *maximized* over points in the feasible region, then we can expect to find a point that is, intuitively, as "far away as possible from the boundary." More precisely, let us seek the solution of the following problem:

$$\text{maximize}_{x_i \in R}\ \ x_1 x_2 x_3$$

$$\text{s.t. } \mathbf{a}^T \mathbf{x} = b, \ \ \mathbf{x} \geq \mathbf{0}. \tag{7.5}$$

The solution of (7.5) exists and is unique,[6] and it is called the *analytic center* of the feasible region. Let us denote it by $\mathbf{x}_C$. The corresponding value of the original objective is $z_C = \mathbf{c}^T \mathbf{x}_C$.

---

[5]Note that the function defined by summing the variables rather than taking their product would not fit our needs.

[6]These facts can be established, for example, along lines outlined in the paragraph immediately preceding expression (7.7).

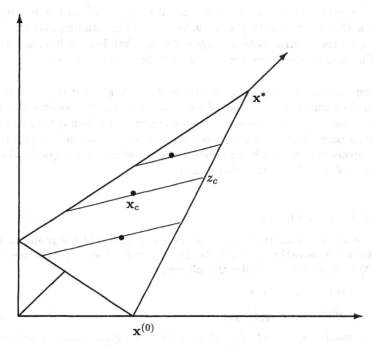

FIGURE 7.2. Points on the central path

Exercise: Use a scaling of variables, say $\bar{x}_j = d_j x_j$, $j = 1, 2, 3$, to transform the linear constraint in (7.5) to $\bar{x}_1 + \bar{x}_2 + \bar{x}_3 = b$. Then use a symmetry argument to solve the problem and obtain an explicit expression for the analytic center.

Let us now consider the contours of the objective function within the feasible region, i.e., points where $c^T x$ has the same value. These are the lines within the triangle shown in Figure 7.2. Let us impose a further restriction and consider only points that lie on one of these contour lines with constant objective value, say z_c. Again, we seek a feasible point that lies as far away from the boundary as possible, as we did earlier when defining the analytic center. This corresponds to solving the problem

$$\text{maximize}_{x_i \in R} \quad x_1 x_2 x_3$$
$$\text{s.t. } a^T x = b, \quad c^T x = z_c, \quad x \geq 0. \tag{7.6}$$

Its solution is a unique point, depicted as x_c in Figure 7.2, which lies on the contour line corresponding to objective value z_c. As this contour line is changed by varying z_c, the associated optimal points x_c trace out a fundamental object of linear programming called the *central path*. Two other interior points on this path are also depicted in Figure 7.2, along with its end points, x^* and $x^{(0)}$, where the original objective is minimized and maximized, respectively.

The central path possesses beautiful properties.[7] In particular, one can show that it is *smooth*, i.e., continuously differentiable, and that the tangent to the central path at any point x_c that lies on it is *parallel to the affine-scaling direction*, computed as in Section 7.1, at x_c.

Computational Exercise: For the specific example given at the beginning of Section 6.1.1 and a contour line of your choice within the feasible triangle (with associated objective value z_c), devise an algorithm for finding the point x_c on the central path. (Hint: It is only necessary to search in one dimension.) Implement the algorithm and run it for different choices of z_c. Give a graphical depiction, akin to Figure 7.2, of the central path.

7.2.2   In General

There are many alternative choices for the objective of the problems (7.5) and (7.6) that achieve the same ends. In the more general setting of the linear program (6.3) in n variables, consider the following:

1. maximize$_{x_i \in R}$ $(x_1 x_2 \ldots x_n)$

2. minimize$_{x_i \in R}$ $\frac{1}{(x_1 x_2 \ldots x_n)}$

3. maximize$_{x_i \in R}$ $(x_1^w x_2^w \ldots x_n^w)$, where w is a given weight, which can be any positive number.

4. minimize$_{x_i \in R}$ $\frac{1}{(x_1^w x_2^w \ldots x_n^w)}$ where again w is a given positive weight.

5. Let P be *any* function of a single variable, say $p \in R$, that is *strictly monotonically increasing* as the variable p increases. Then replace any of the quantities being maximized or minimized within the preceding four objectives by the univariate function P acting on that quantity. For example, take $p = x_1 x_2 \ldots x_n$, and maximize$_{x_i \in R}$ $P(x_1 x_2 \ldots x_n)$.

For the example of Section 7.2.1, identical solutions defining the analytic center and central path would be obtained from (7.5) and (7.6) when any one of the five objectives itemized above is used instead.[8] To generalize these two defining problems to the linear program[9] in standard form (6.3) and arbitrary dimensions m and n, replace the single constraint $a^T x = b$ in (7.5) and (7.6) by the m constraints $Ax = b$, and the objective by any one of the five itemized above.

Exercise: Explain why the alternative choice of objective function $\sum_{i=1}^{3} x_i$ would *not* be suitable for defining an analytic center for (6.1).

---

[7] For additional detail on establishing properties of the central path, see also the discussion in Section 7.3.

[8] The reasons for this equivalence will become clearer in the discussion leading up to expression (7.7).

[9] Recall our previous assumption that the feasible region is *bounded*, which is not necessary but is made for ease of discussion, and that its interior is nonempty.

Exercise: Explore the use of different weights for components x_i in items 3 and 4 above. For example, consider an objective of the form maximize$_{x_i \in R} \prod_{i=1}^{n} x_i^{w_i}$, where $w_i, i = 1, \ldots, n$, are positive numbers. Define the associated *weighted* analytic center and *weighted* central path.

Exercise: Show that the objective in (7.5) and (7.6) could be replaced by

$$\text{minimize}_{x_i \in R} \sum_{i=1}^{3} \frac{1}{x_i},$$

but in this case the analogously defined analytic center and path of centers would differ from the ones obtained from (7.5) and (7.6). Also define a weighted analytic center and weighted central path as in the previous exercise.

Exercise: In the itemized list of objectives above, show that items 2, 3, and 4 are special cases of item 5; i.e., they correspond to particular choices of the monotonic function P.

One particular choice of the function P in item 5 above has *overshadowed* all others, namely, $P = \ln$, the natural logarithm. This can be applied to any of the four objectives defined in items 1 through 4 above. Note that the logarithmic function $\ln(x_i)$ of the variable x_i goes to $-\infty$ as x_i approaches its value 0 on the boundary of the feasible region, and it increases monotonically as x_i increases. By using the expression $\ln\left(\prod_{i=1}^{n} x_i^w\right) = w \sum_{i=1}^{n} \ln(x_i)$ in conjunction with two well-known facts—minimizing a function is equivalent to maximizing its negative; the solution of an optimization problem is unchanged when its objective function is multiplied by a positive constant—it is easy to verify that one obtains identical *problems* defining the analytic center and central path (not merely identical solutions) for each of the four cases. Furthermore, the logarithmic transformation is very useful for characterizing the solutions of these problems, for example, demonstrating existence and uniqueness of solutions for a particular objective, or equivalence of solutions for the objectives itemized above.

Thus, *the following problem*[10] *for defining the analytic center is accepted nowadays as the norm:*

$$\text{maximize} \sum_{i=1}^{n} \ln(x_i)$$

$$\text{s.t. } \mathbf{Ax = b}, \quad \mathbf{x > 0}, \tag{7.7}$$

where we use ">" in the bounds on variables, because the logarithmic function is not defined at $x_i = 0$. Again denote the unique solution by $\mathbf{x}_C$ and the associated objective value by $z_C = \mathbf{c}^T \mathbf{x}_C$.

To define the central path using an approach similar to that of the previous subsection 7.2.1, consider the contours of the objective function ("hyperplanes," instead of lines, on which $\mathbf{c}^T \mathbf{x}$ has the same value) within the feasible region.

---

[10] Its objective is an example of a "concave" function, the negative of a "convex," or "bowl-shaped," function; and in consequence the earlier function $x_1 x_2 \ldots x_n$ from which this objective was derived is called "log-concave."

Restrict attention to points that lie on one of these contours corresponding to objective value z_c, and again seek a feasible point as "far away" as possible from the boundary, by solving the following problem:

$$\text{maximize } q = \sum_{i=1}^{n} \ln(x_i)$$

$$\text{s.t. } \mathbf{Ax} = \mathbf{b}, \quad \mathbf{c}^T \mathbf{x} = z_c, \quad \mathbf{x} > \mathbf{0}. \tag{7.8}$$

Its solution is a unique point, say $\mathbf{x}_c$, on the contour corresponding to objective value z_c. As this contour is changed by varying z_c, the optimal points $\mathbf{x}_c$ trace out the *central path* of (6.3). The path's properties mentioned at the end of Section 7.2.1 remain true, of course, in the more general setting considered in the present section.

Finally, (7.8) provides an alternative way to define the analytic center. Observe that the value q of its objective function changes as z_c is varied. The analytic center x_C is the point where q itself is as large as possible. In other words, the analytic center of the feasible region is the solution of the following problem, which now treats z_c as a variable:

$$\text{maximize}_{z_c \in R} \left[\max_{x_i \in R} \sum_{i=1}^{n} \ln(x_i) \text{ s.t. } \mathbf{Ax} = \mathbf{b}, \quad \mathbf{c}^T \mathbf{x} = z_c, \quad \mathbf{x} > \mathbf{0} \right]. \tag{7.9}$$

7.3   Interior-Point Algorithms

7.3.1   Example in 3-D

Our goal has been to solve the linear program (6.1). When its constraints are retained but its linear objective $\mathbf{c}^T \mathbf{x}$ is replaced by the (nonlinear) objective of maximizing the quantity $x_1 x_2 x_3$, or equivalently *minimizing* the quantity $-x_1 x_2 x_3$, then the resulting problem (7.5) has the analytic center of the feasible region as its solution. Suppose we wanted to build a "path" that joins the analytic center to the solution of the linear program. Then it would be natural to define a new objective as a *weighted combination* of the previous two objectives, namely, $\mathbf{c}^T \mathbf{x} - \mu \, (x_1 x_2 x_3)$, where $\mu \geq 0$ is a parameter, and to minimize this objective subject to the original linear constraint and the nonnegative bounds on variables. For subsequent purposes, let us also define

$$C(\mathbf{x}) = x_1 x_2 x_3. \tag{7.10}$$

Thus, let us consider the problem

$$\text{minimize } \mathbf{c}^T \mathbf{x} - \mu \, C(\mathbf{x})$$

$$\text{s.t. } \mathbf{a}^T \mathbf{x} = b, \quad \mathbf{x} \geq \mathbf{0}, \tag{7.11}$$

where μ is any nonnegative real number. When $\mu = 0$ then (7.11) is the original linear program. As μ approaches infinity (denoted henceforth by ∞), the second term dominates, and the solution of (7.11) approaches the analytic center. For each value of μ, the solution of (7.11) can be shown to occur at a *unique* point, which we denote by $\mathbf{x}(\mu)$. What do the solution points $\mathbf{x}(\mu)$ define as the parameter μ is reduced monotonically from ∞ to 0? *The answer*, which the reader will likely have guessed, *is that they trace the segment of the central path of the linear program from the analytic center, $\mathbf{x}_C$, to the optimal solution, $\mathbf{x}^*$!*

To verify this result, consider two nonnegative values of the parameter, say μ_1 and μ_2, such that $\mu_1 > \mu_2$. Since $\mathbf{x}(\mu_1)$ is the unique optimal point of (7.11) with $\mu = \mu_1$, and $\mathbf{x}(\mu_2)$ is feasible for its constraints, it follows that

$$\mathbf{c}^T\mathbf{x}(\mu_1) - \mu_1 C(\mathbf{x}(\mu_1)) < \mathbf{c}^T\mathbf{x}(\mu_2) - \mu_1 C(\mathbf{x}(\mu_2)).$$

Similarly,

$$\mathbf{c}^T\mathbf{x}(\mu_2) - \mu_2 C(\mathbf{x}(\mu_2)) < \mathbf{c}^T\mathbf{x}(\mu_1) - \mu_2 C(\mathbf{x}(\mu_1)).$$

The foregoing inequalities, in turn, imply that

$$\mu_1\left[C(\mathbf{x}(\mu_1)) - C(\mathbf{x}(\mu_2))\right] > \mathbf{c}^T\mathbf{x}(\mu_1) - \mathbf{c}^T\mathbf{x}(\mu_2) > \mu_2\left[C(\mathbf{x}(\mu_1)) - C(\mathbf{x}(\mu_2))\right]. \tag{7.12}$$

There are two cases to consider:

(a) Suppose $[C(\mathbf{x}(\mu_1)) - C(\mathbf{x}(\mu_2))] \geq 0$. Then $\mathbf{c}^T\mathbf{x}(\mu_1) > \mathbf{c}^T\mathbf{x}(\mu_2)$.
(b) Suppose, on the other hand, $[C(\mathbf{x}(\mu_1)) - C(\mathbf{x}(\mu_2))] < 0$. Then multiply (7.12) by -1, which reverses the inequalities, yielding

$$\mu_1\left(C(\mathbf{x}(\mu_2)) - C(\mathbf{x}(\mu_1))\right) < \mathbf{c}^T\mathbf{x}(\mu_2) - \mathbf{c}^T\mathbf{x}(\mu_1) < \mu_2\left(C(\mathbf{x}(\mu_2)) - C(\mathbf{x}(\mu_1))\right). \tag{7.13}$$

The quantities in parentheses are positive, and one obtains an immediate contradiction to $\mu_1 > \mu_2$. The premise for (b) must therefore be false.

Thus, the value of the original objective function at the solution of (7.11), i.e., the quantity $z(\mu) = \mathbf{c}^T\mathbf{x}(\mu)$, decreases monotonically from z_C, as μ decreases from ∞, where z_C is defined in Section 7.2.1. Furthermore, suppose the value of the original objective function $\mathbf{c}^T\mathbf{x}$, i.e., the first term in the objective of (7.11), is kept fixed at a contour value $z(\mu)$ that lies between z_C and the optimal value of the linear program. Then the problem (7.11) is equivalent to (7.6) after one identifies $z(\mu)$ with z_c, and it has solution $\mathbf{x}(\mu) \equiv \mathbf{x}_c$. These observations, in turn, imply that the points $\mathbf{x}(\mu)$ must trace the central path from the analytic center to the optimal solution.

Exercise: If the sign of the original objective function is reversed, i.e., $\mathbf{c}$ is replaced by $-\mathbf{c}$, verify that the other segment of the central path, leading from the analytic center to the *maximizing* point of the original objective, is traced by the solution of (7.11), as μ is decreased from ∞ to 0.

The problem (7.11) is a specialized *nonlinear* program. Its objective is a particular instance of a so-called posynomial function—other examples are itemized at the beginning of Section 7.2.2—and the problem (7.11) itself is a particular instance of a so-called geometric program. The foregoing development provides one illustration of the increasingly close alliance that developed, post-1984, between linear programming and the wider subject of nonlinear programming. This was perhaps the most far-reaching consequence of the Karmarkar revolution. Over the course of time, it led to a gradual supplanting of the particular building blocks employed by Karmarkar [1984] to achieve his breakthrough. The new bridges that were built between linear programming (LP) and nonlinear programming (NLP), and the revitalization of certain NLP techniques that had fallen into disuse, created a much more unified subject and a wealth of new NLP-inspired algorithmic approaches for solving a linear program. We will outline these advances in the next section, and continue with their discussion in the following chapter, Section 8.6.

7.3.2   In General

Karmarkar [1984] used a special form of the standard linear program (6.3) and so-called projective-scaling transformations of it. These techniques motivated simpler *affine-scaling* transformations applied directly to (6.3), as described in Section 7.1. It was then discovered that affine scaling and a mathematical algorithm for solving linear programs based on this idea had been proposed and analyzed much earlier by the Russian mathematician Dikin [1967]. However, Dikin's algorithm had been all but forgotten. It reemerged as a viable *numerical algorithm*[11] after it was independently discovered by several researchers as a simplified (and often just as effective in practice) variant of the algorithm of Karmarkar [1984]. The recognition of the affine-scaling algorithm as a *practical* numerical technique for linear programming was one of the early discoveries of the post-Karmarkar period.

Centering, the other key idea discussed in this chapter, was to some extent implicit in Karmarkar's breakthrough. But the discoveries of the *analytic center* and the *central path*, two fundamental objects of algorithmic linear programming, were due to Sonnevend [1986], Megiddo [1989], and others, with antecedents that were subsequently found in Huard [1967] and McLinden [1980]. As we have seen, the techniques that can be used to define the analytic center and central path are related to so-called geometric programming and underlying posynomial functions; see the notes at the end of this chapter for a reference on these topics. Let us now expand on the development in Section 7.3.1 by making the following key observation:

Proposition: The central path characterization defined by (7.11) and the argument immediately following it (given in small print) will remain unaltered when

---

[11]See Section 6.2 for a discussion of this term.

the function $C(\mathbf{x})$ in (7.10) is replaced by *any* one of the functions itemized at the beginning of Section 7.2.2 with the appropriate sign: negative when maximizing and positive when minimizing. Also, in order to match dimensions in (7.10), take $n = 3$.

We leave the verification of this proposition as a straightforward exercise for the reader. It would obviously also be valid for arbitrary dimension n.

Let us now make the specific choice $P = \ln$ in item 5 of Section 7.2.2. For the linear program in standard form (6.3), the problem (7.11) generalizes[12] as follows (see also (7.7)):

$$\text{minimize } \mathbf{c}^T \mathbf{x} - \mu \sum_{i=1}^{n} \ln(x_i)$$

$$\text{s.t. } \mathbf{Ax} = \mathbf{b}, \quad \mathbf{x} > \mathbf{0}, \tag{7.14}$$

The foregoing "log-barrier transformation," which is an alternative way to characterize the segment of the central path leading from the analytic center to an optimal solution, is a well-known technique of nonlinear programming with roots in the work of the Nobel laureate R. Frisch [1955], [1956];[13] see also the definitive monograph of Fiacco and McCormick [1968; 1990]. It was quickly recognized to be closely related to Karmarkar's "potential function," the third key idea used within his polynomial-time algorithm. Thus, classical techniques of nonlinear programming[14] and algorithms based on them were discovered to have remarkable and hitherto unsuspected polynomial-time complexity properties in the setting of linear (and more generally convex) programming. They became invaluable for establishing properties of the central path, for example, the ones described at the end of Section 7.2.1. And most importantly, they were found to be efficient for solving large-scale linear programs in practice, thus giving birth to a new cycle of algorithmic and software development in optimization.

---

[12]This is an example of a convex program; see also Section 8.6.

[13]In the title of this seldom-referenced article in French, the operative terms "linear programming," "logarithmic," and "potential" can all be found in translation.

[14]During the years prior to 1984, the logarithmic barrier method had fallen into disuse and increasingly became viewed as an outmoded technique of nonlinear programming. The fact that it possessed a catalog of remarkable properties *in the setting of large-scale linear programming* was discovered during the decade of intense activity, by researchers worldwide, that followed Karmarkar's breakthrough. A conventional log-barrier viewpoint of "pushing away from a constraint boundary" was supplanted by the notion of "hewing to a path of centers," tightly in order to obtain polynomial-time convergence, or very loosely when the goal was good practical performance. Historical parallels for such a shift in perspective can be found in Dantzig's account of his discovery of the simplex algorithm; see his illuminating remarks quoted in Section 6.2. When viewed conventionally in the "row geometry" of a linear program, it was initially rejected, on intuitive grounds, as being unpromising. But in an alternative "column geometry," the simplex algorithm was revealed to be potentially very efficient.

7.4 Notes

Sec. 7.1: This material is derived from Nazareth [2003, Chapter 7], where a numerical illustration of the performance of the affine-scaling algorithm vis-à-vis the simplex algorithm can also be found.

Secs. 7.2-7.3: The central path and interior-point formulations continue the line of development in Nazareth [2003, Chapter 7]. An excellent discussion of posynomials, geometric programs, and their "log-convexity" properties can be found in Avriel [1976]. For a definitive treatment of the topics discussed in this chapter, see Nesterov and Nemirovskii [1994], Renegar [2001], and references cited therein.

8
Nonlinear Programming

As we have seen in previous chapters, an optimization problem typically involves maximizing or minimizing an objective function over a feasible region defined by a set of constraints (including bounds on variables). Whenever the objective and/or constraints do not match the linear programming definition of Chapters 5–7, then the optimization problem is, by default, a *nonlinear programming problem*. We now give a brief overview of problems of this type.

Our focus will be on problems involving a *finite* number of variables that assume *continuous*, real values, and for which the data defining constraints and objectives are *deterministic*, or known with certainty. Other more general types of optimization problems, for example, problems where variables are explicitly restricted to discrete values, typically integers, or problems where the data defining constraints and/or objectives are known only probabilistically, will be addressed in Chapter 10.

In Sections 8.1 and 8.2, nonlinear unconstrained optimization is introduced from geometric and algebraic viewpoints on two-dimensional problems. Section 8.3 addresses the key issue of the cost of problem information. Unconstrained problems of arbitrary dimension and the inclusion of constraints are the topics of Sections 8.4 and 8.5, respectively. Finally, Section 8.6 provides a schematic overview of the main problem areas of nonlinear programming, in particular, for the case in which objectives and constraints are defined by smooth, or differentiable, functions.

8.1   Geometric Perspective

Let us return to the example of Section 1.4. A point is sought on the surface of an opaque lake where the water is deepest. Two coordinates in, say, the E-W and N-S directions are needed in order to specify the location of a point on the surface. The water depth at each such point (x, y) is the vertical distance beneath the surface to the lake's bottom. Thus let us define the depth function value $d(x, y)$ to be the *negative* of this distance, i.e., if the distance measured by the length of rope attached to an anchor dropped at (x, y) is 10 units (for example, meters), then $d(x, y)$ is -10. For a point on the shoreline, the depth function value is 0. We seek the point on the surface, say (x^*, y^*), such that $d(x^*, y^*)$ is as negative as possible. In other words, we seek to *minimize* the depth function value.

In order to obtain a clearer description of this optimization problem, let us hypothesize a gradual draining of the lake by pumping out the water. The initial shoreline of the lake will be called the *contour* corresponding to depth function value zero, and the initial surface of the lake will be called the associated *level set* (also for depth function value zero). As the water is gradually pumped out, new shoreline contours and level sets will become visible, corresponding to depth function values that *decrease* continuously from zero. And as the water level falls, the shape of such contours and their associated level sets changes, perhaps dramatically, depending on the topography of the underwater landscape (the "bottomscape").

Two situations can arise. Firstly, the water within each level set may remain *contiguous throughout* the lowering of the surface, and eventually all the water will be drained from the lake and the coordinates defining the location on the original surface of its deepest point will be revealed. This point is the unique *global* minimum of the lake. (Again speaking hypothetically, the entire lake could therefore be drained by gravity from a single, strategically placed outlet.) Alternatively, as drainage proceeds, level sets may become disconnected as water settles into "ponds" that are isolated from one another. Each pond must then be drained individually, and will eventually reveal a *local* optimizing point (or separate into even more ponds). The deepest one among them (relative to the initial surface of the lake) is again called the global optimum. Locating it is a harder task than the earlier case in which the level sets exhibited a "contiguity" property, and the global minimum is a unique local minimum.

Of course, drainage of the lake is not an option when the optimization problem posed in Section 1.4 must be solved. The contours revealed above must remain invisible, and only *individual points on particular contours* are "sounded" by the boatman, who drops anchor to obtain the water depth, and optionally, the slope of the underwater terrain and its curvature.[1] Many

---

[1] Assume, hypothetically, that the boatman's anchor has sensors that are able

different families of algorithms have been proposed for using information available in this limited fashion to solve the underlying optimization problem.

For example, algorithms of so-called Newton–Cauchy type proceed as follows: begin at some initial point, find the depth/slope/curvature, and from this information determine a *direction of descent*, i.e., a direction of movement *on the surface* along which the depth function value initially decreases. Travel along this direction, taking additional soundings as necessary, in order to obtain a point with a lower depth function value. This is called a "line search." At the new point, or "current iterate," determine a fresh direction of descent from information available there and at any (or all) prior iterates. Initiate another line search, and repeat the procedure until it converges. Individual algorithms in this family differ in the information that is gathered, the way information is used to obtain a descent direction, the techniques used within the line search, and the computer storage requirements of the algorithm (full memory vis-à-vis limited memory).

Direct search algorithms provide another option for the boatman. They are based on an organized sampling of points using some systematic pattern or alternatively a statistical, or random, procedure. Only the depth is evaluated, i.e., the algorithms in this family do not employ slopes or curvature. For instance, in the Nelder–Mead procedure, points at the vertices of a triangle on the surface are sounded for water depth. The *worst point* (largest depth function value[2] in *magnitude* among the three vertices) and the *mid-point* of the other two vertices are computed. Along the line joining these two points, one or more more additional samples are taken, at fresh points that again fall into some specified pattern. The triangle is updated by replacing the worst vertex by one of the new points generated. Then the procedure is repeated. Algorithms of this direct-search type can often step over local minima and are thus able to cope with "noise" in the depth function values.

Several other families of algorithms for unconstrained minimization enjoy widespread use: so-called conjugate gradient methods, trust-region algorithms, simulated annealing, evolutionary algorithms, and so on. For details, see the references cited in the notes at the end of this chapter.

---

to measure the slope (rate of change of depth along the coordinate axes and hence the direction of most rapid rate of change, or *gradient* vector) and the curvature (information that characterizes the rate of change of the slope itself, formally the *Hessian* matrix).

[2]If there are ties, pick any among them.

8.2 Algebraic Perspective

In nonlinear unconstrained minimization, a point is sought that minimizes an objective function. The latter is often defined explicitly by an algebraic expression of the problem variables. For example, the topography of the lake bottom in the previous section could be defined by the following function of two variables, which correspond to the coordinates, say x and y, of a point on the surface:

$$f(x, y) = 1 - 2x + x^2(1 + 100x^2) + 100y(y - 2x^2). \qquad (8.1)$$

We will discuss the correspondence between $f(x, y)$ and the depth function $d(x, y)$ in the next subsection, and for the moment, let us simply consider the problem of finding a (local) minimizing point of (8.1).

The discernible features of the undulating landscape defined by (8.1) are not immediately evident, but it is possible to reformulate this function and reveal these features as follows:

$$f(x, y) = (1 - x)^2 + 100(y - x^2)^2. \qquad (8.2)$$

In this form, it is recognizable as the so-called Rosenbrock's function,[3] which is widely used as a test function in the optimization literature. We see immediately from (8.2) that the function always has a nonnegative value and that its minimizing point is at $(1, 1)$. Observe that the second term defines a parabola on the coordinate (x, y) plane, and the value of the term $(y - x^2)$ at points that lie off this parabola are amplified by a large constant 100. This produces the prominent feature of the landscape defined by the function, namely, a steep, curving valley and, in particular, a topography that is challenging to traverse in a neighbourhood of the origin.

Contours and level sets for Rosenbrock's function are defined as in the previous section. Thus, the contour that corresponds to the function value v is the set of points for which $f(x, y) = v$, and the corresponding *level set* of f is the set of points such that $f(x, y) \le v$. Rosenbrock's function is very inexpensive to evaluate on a computer, and thus it is possible to plot its contours. This requires many evaluations of the function and is the analogue of draining the lake as described in the previous section.

Computational Exercise: Use MATLAB or a similar computer language of your choice to develop contour plots of Rosenbrock's function.

It is easy to obtain some idea of the contours by inspection, without plotting them by computer. Consider a contour corresponding to $v = 121$.

---

[3]Rosenbrock's function is normally specified as a sum of squares as in (8.2). Its disguised form (8.1) highlights the fact that one cannot find a minimizing point by inspection in the more typical case; i.e., an optimization algorithm must be employed.

The points (x, y) that lie both on the above curving parabola and this contour correspond to the solution of $(1 - x)^2 = 121$, namely, $(-10, 100)$ and $(12, 144)$. Points on the contour that lie on the (vertical) y-axis, where the x component is zero, are given by the solution of $1 + 100y^2 = 121$, i.e., $(0, \sqrt{120}/10)$ and $(0, -\sqrt{120}/10)$. A "banana-shaped" contour will pass through these four points. The contours will turn to ellipsoidal shapes around the minimizing point $(1, 1)$. The associated level sets, as v is reduced from our assumed intial value of 121, have the "connectedness" property mentioned in the previous section,[4] and thus the point $(1, 1)$ is a global minimum.

Algorithms for minimizing Rosenbrock's function would proceed just as described at the end of the previous section, but now the requisite information[5] is computed from f. See again the references cited in the notes at the end of this chapter for further detail.

8.3   Information Costs

An oft-used algorithmic metaphor for optimizing an unconstrained function is that of a marble rolling downhill to a minimizing point on a mountainous, Rosenbrock-like landscape. However, this metaphor is misleading. On its downward path, a marble samples the landscape for height, slope and curvature, continuously and at no expense. The metaphor of a small boat floating on an *opaque* lake, whose "bottomscape" is defined by a Rosenbrock-like function as in the example of Section 8.1, is preferable, because it highlights a central tenet of algorithmic optimization: the acquisition of information at any point incurs *a significant, nonzero cost*. In practical applications, this cost often outweighs *all* other costs associated with manipulating information within an unconstrained minimization algorithm.

To obtain a correspondence between the geometric perspective based on the boat on a lake metaphor and the algebraic perspective of the previous section, let $(x^{(0)}, y^{(0)})$ be any starting point used to initiate a minimizing algorithm for f, and then define the depth function d as follows:

$$d(x, y) = f(x, y) - f(x^{(0)}, y^{(0)}). \tag{8.3}$$

This depth function has value zero on the "shoreline" contour corresponding to the point $(x^{(0)}, y^{(0)})$, and it is negative for all points within the corresponding level set. On a computer, $d(x, y)$ is very inexpensive to compute. But one can *simulate* the earlier "anchor dropping" situation, where

---

[4]Formally this is related to the mathematical notion of "lower semicontinuity."

[5]For example, in a Newton–Cauchy algorithm, the slope, or formally the gradient vector of f, and the curvature, or formally the Hessian matrix, can be obtained from the first and second partial derivatives of f.

information is costly, by including an expensive-to-compute constant as follows:

$$d(x,y) = f(x,y) - f(x^{(0)}, y^{(0)}) + \Theta(T), \qquad (8.4)$$

where $\Theta(T)$ is a function that returns any prespecified *constant* value. It just moves the function defined by (8.3) up or down by a fixed amount and does not alter its minimizing point, but it incurs T additional units of computational cost, where T can be chosen as large as desired. (Within a function evaluation subroutine, one could, for instance, simply *recompute* $f(x,y)$ an arbitrarily large number of times in a loop before returning its value.) In an actual simulation, there is no need to compute $d(x,y)$ from (8.4). Instead, just compute $f(x,y)$ and count the *number of calls* to its evaluation subroutine. This count itself can be used as a measure of computational cost of an optimization algorithm—each call to the subroutine can then be viewed as *implicitly* incurring T units of computational cost. Rosenbrock's function is often used in this manner as a test problem to *simulate* a more realistic situation, where a function may have Rosenbrock-*like* contours but is very expensive to evaluate on a computer. A large body of test problems are available in the optimization literature and are used to compare algorithms in this manner.

8.4   Dimensions

The contours of functions in up to three variables can be visualized. But for dimensions that exceed 3, geometric intuition is lost and one must resort to the algebraic perspective.

Unconstrained optimization problems of arbitrary dimensionality can be divided into three broad categories:

- Unidimensional problems, i.e., problems involving a single variable. These are distinctive and are solved by very specialized algorithms; see also Section 8.6.

- Problems in a modest numbers of variables. These are typified by the examples considered earlier, for instance, Rosenbrock's function in two dimensions.

- Large-scale problems, i.e., problems involving many variables. For example, Rosenbrock's function can be extended to an arbitrary number of dimensions, or variables $x_1, \ldots, x_n$, as follows: group the variables in pairs, $(x_i, x_n), i = 1, \ldots, n - 1$, and repeatedly use the expression (8.2) with the identification $x \equiv x_i$ and $y \equiv x_n$. Then sum the $n - 1$ terms, yielding the function

$$f(x_1, \ldots, x_n) = \sum_{i=1}^{n-1} \left[(1 - x_i)^2 + 100(x_n - x_i^2)^2 \right], \qquad (8.5)$$

for an arbitrarily large choice of integer n.

The algorithms of Section 8.1 all have natural counterparts in higher dimensions. It is important to note that a low-dimensional unconstrained problem, i.e., a problem with a small number of variables, is *not* necessarily easier to solve than a problem of higher dimension. For example, minimizing a strictly convex quadratic function—a function whose countours are "hyperellipsoids"—in n variables can be done simply by solving an associated system of linear equations involving a single $n \times n$ symmetric matrix,[6] which is an inexpensive operation. In comparison, a two-dimensional function that is Rosenbrock-like, but with *high informational costs*, as in Section 8.3, may be a much more challenging problem.

8.5   Constraints

Constraints can be placed on the region of search. In the example of Section 8.1, place a physical boom that constrains the boatman's search to a circular region of the lake, or place several linear booms that define a polygonal region. Suppose the original minimizing point lies outside the feasible region defined by these physical constraints. Again imagine pumping out the water. At some stage of drainage, the current level set will lie entirely *outside* the feasible region, but touch its boundary. This point of contact is then the minimizing point of the *constrained* problem.

In the algebraic perspective, the circular feasible region centered on a point, say, (p, q), and of radius, say, r, would be defined by the nonlinear constraint $(x - p)^2 + (y - q)^2 \leq r^2$. The polygonal feasible region would be defined by a set of linear constraints involving two variables, for example, $ax + by \leq c$, where $a, b,$ and c are real numbers Alternatively, the boat might be constrained by equality constraints. In the foregoing examples, replace "$\leq$" by "$=$." In general, an optimization problem in two dimensions would be defined by minimizing a function $f(x, y)$ subject to a set of equality and/or inequality constraints, and bounds on the variables x and y.

It is worth noting that the inclusion of constraints does *not* always increase the difficulty of solving the optimization problem. For example, suppose one seeks to minimize Rosenbrock's function within a small square centered at the point $(1, 1)$, with sides of length say 0.5. Then the optimization task is facilitated, because most of the steep curving valley, in particular, the region around the origin, is excluded. Within the square feasible region, the contours of the function are bowl shaped and *resemble* those of a strictly convex quadratic function (see Section 8.4). This is an

---

[6]In particular, a strictly convex quadratic function in two variables has ellipsoidal contours and can be solved by inverting an associated 2×2 symmetric matrix.

"easy" optimization problem. In other words, constraints may complicate the formulation of a suitable optimization *algorithm*, but they could, in fact, ease the optimization problem-solving *task*.

As in the previous section, equality, inequality, and bound constraints can be defined over an arbitrary number of variables. Linear constraints over n variables have been discussed in Chapters 5–7. The foregoing circular feasible region would extend to a so-called hypersphere. When centered at a point $(p_1, \ldots, p_n)$, with radius r, it would be defined by the nonlinear inequality constraint

$$\sum_{i=1}^{n} (x_i - p_i)^2 \le r^2. \tag{8.6}$$

An example of a nonlinear optimization problem would then be given by minimizing the objective function (8.5) subject to the constraint (8.6).

8.6   Differentiable Programming

An optimization problem is defined, in general, by minimizing an objective function in n variables subject to a set of equality and inequality constraints, along with bounds on the variables. The functions defining objectives and constraints in the area traditionally designated as "nonlinear programming" are assumed to be smooth. Since nonlinear equations are instances of equality constraints, the subjects of nonlinear optimization and nonlinear equation solving have traditionally shared a common frontier. For reasons discussed in Chapter 7, nonlinear programming is nowadays also closely integrated with linear programming. Thus, differentiable optimization and equation solving, or more compactly, *differentiable programming*, is a preferable name for the unified field. This name also provides a natural counterpart to the area of *nondifferentiable programming*, where the functions defining objectives and constraints are permitted to be nonsmooth.

Feasible regions defined by given constraints and bounds on variables fall into two broad categories called *convex* and *nonconvex*, respectively. We will say that a feasible region is a *convex set* provided every point of the set is in the "line of sight" of every other point; i.e., given *any* two distinct points of a convex set, the entire line joining the two points also lies in the set. Otherwise, the region is said to be nonconvex.

Exercise: Show that the feasible region defined by the earlier nonlinear constraint $(x - p)^2 + (y - q)^2 \le r^2$ is a convex set, and the feasible region defined by the equality constraint $(x - p)^2 + (y - q)^2 = r^2$ is nonconvex. Are the corresponding statements true when the constraint is linear?

Computational Exercise: Impose a feasible region on the contour plot of the previous MATLAB exercise of Section 8.2 and identify the optimal solution graphically. Consider both convex and nonconvex feasible regions.

Optimizing an arbitrary objective function f over a convex feasible region is

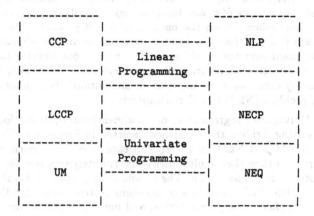

FIGURE 8.1. Differentiable Programming

a simpler task than optimizing it over a region that is not necessarily convex. Convexity implies that the feasible region is contiguous and permits movement in straight lines, or so-called feasible directions, from each point to *any* other point within the set. And it facilitates the use of geometric operations like "projection onto a feasible set" within the design of algorithms; see Section 7.1.2 for an example of projection.

The various models of differentiable programming are displayed in the schematic of Figure 8.1, and the acronyms used within it are defined below. The *watershed* in the figure is between problems whose feasible regions are convex and problems whose feasible regions are not necessarily convex.

- UM, LCCP, CCP: Algorithms for unconstrained minimization (UM) seek a minimizing point, i.e., a point of local convexity, of an arbitrary smooth objective function over the unrestricted space of variables $(x_1, \ldots, x_n)$, denoted by R^n. The latter is obviously a convex region. The linear convex-constrained programming problem (LCCP) and the convex-constrained programming problem (CCP) are natural extensions, where R^n is replaced by a convex polytope (set defined by linear equalities and inequalities) and a more general convex region, respectively. When, additionally, the objective function is a convex function,[7] then the optimization problem is said to be a *convex programming problem* (CP). We have already encountered an example in (7.14) of the previous chapter.

- NEQ, NECP, NLP: Points that satisfy a set of nonlinear equations (equal-

---

[7]A *sufficient* condition for f to be a convex function is that all its level sets are convex. For example, Rosenbrock's function is not a convex function over its entire domain. It is (locally) convex when restricted to a region around its minimizing point, where the level sets are ellipsoidal.

ity constraints) form a set that is *not* necessarily convex. Thus, the nonlinear equation-solving problem (NEQ), where n nonlinear equations are solved over a space of n variables, belongs naturally on the nonconvex side of the watershed. When the number of equality constraints in the NEQ problem is fewer than the number of variables and an objective function is optimized over the resulting feasible region, one obtains the nonlinear equality-constrained programming problem (NECP). When, in addition, inequality constraints are introduced, one obtains the nonlinear programming problem (NLP) in full generality.

- 1-D: Univariate Programming, or unidimensional optimization and equation solving, bridges the watershed depicted in Figure 8.1. Algorithms for solving UM and NEQ for the special case $n = 1$ are much more closely interrelated than their multidimensional counterparts; see Nazareth [2003; Chapter 4] for a discussion. The classic monographs of Traub [1964] and Brent [1973; 2002] provide a comprehensive treatment of 1-D algorithms and their effective implementation, and both remain relevant today.

- LP: Linear programming is the other bridge that traverses the convexity/nonconvexity watershed. Some of its key algorithms are derived, in a natural way, by approaching the subject from the convexity side, as we have seen in Chapter 7. Other so-called primal-dual algorithms approach linear programming from the opposite side of the watershed, where primal and dual variables are treated simultaneously, often as equal partners; further details can be found in Nazareth [2003, Chapter 10].

Again, see Nazareth [2003] for additional discussion of Figure 8.1 and a detailed study of algorithms for solving the UM, NEQ, 1-D, and LP problems, which lie at the foundations of differentiable programming; and see Bertsekas [1999] for a comprehensive treatment of the subject (including the LCCP, CCP, NECP, and NLP problems) that also conforms to the overall pattern of Figure 8.1.

8.7 Notes

Section 8.6: The so-called Karush–Kuhn–Tucker (KKT) *optimality conditions* for differentiable programming problems on the convex side of the watershed can be formulated in an analogous manner to that of Section 5.2.3; see Zangwill [1969] for a concise and very readable account. For the (technically more difficult) derivation of the *same* conditions for problems on the nonconvexity side of the watershed, see the very accessible account in Luenberger [1984]. The three books cited above—Zangwill [1969], Luenberger [1984], and Bertsekas [1999]—form a natural progression, each inspired by its predecessor, and coincidentally spaced at 15-year intervals.

9

D$_L$P and Extensions

Optimization problems that arise in practice are often *dynamic* in nature; i.e., they are defined over a time axis, and instead of conforming to one of the "pure" optimization models of earlier chapters, they may exhibit a *mix* of network, linear, and/or nonlinear programming characteristics.

In this chapter we consider problems of this type. Specifically, we describe optimization models that are defined on a network of states of a discrete dynamical system,[1] in conjunction with linear programming constraints and objectives. Henceforth, they will be identified by the acronym D$_L$P. They are useful for capturing a wide variety of natural and renewable resource-planning optimization problems. We will describe two simple D$_L$P examples, involving timber and rangeland resources.

Computer software is available that greatly facilitates experimentation with D$_L$P optimization models. It provides a useful educational tool as described in the concluding section.

9.1   A Timber-Harvesting Problem

Let us return to the third example of Chapter 1 (Section 1.3), which is derived from a forestry project in southern Tanzania. Its forest planners

---

[1]See the last item of the list of network application areas in the introduction to Chapter 2.

address[2] the problem of

> $\cdots$ deciding on the optimal thinning schedule to adopt for *Pinus patula* stands grown on a 30-year rotation in order to maximize total physical yield from the stands over the rotation period. At present no thinning is being done except in some experimental compartments. However, the project manager wants to evaluate the possibility of thinning when the stands are 10 years old and again at 20 years. The purpose of thinning would be twofold: (1) to harvest small and diseased trees that would normally die before the stand reaches rotation age and whose volume would thus be lost without thinning and (2) to provide more light and nutrients for the trees that remain after thinning so that diameter growth (and log value) would be accelerated. Like most foresters, the manager of Sao Hill Misitu wants to simplify the instructions given to the work crews as much as possible. Thus, for each potential thinning age the manager plans to consider only three possibilities: (1) no thinning at all, (2) a light thinning to remove about 20 percent of the standing volume, and (3) a heavy thinning to remove about 35 percent of the standing volume.

Initially, the stand is of age 10, and the average standing volume is 260 cubic meters per hectare. Figure 9.1 summarizes the set of possible planning alternatives.

Each node of the network corresponds to a state of the forest resource and is defined by the age of standing timber (in years) and the average volume of standing timber (in cubic meters per hectare). For example, the first node S_1 is defined by age 10, volume 260. The quantity associated with each arc (transformation from one state to another) gives the volume of timber that is removed at the start of the ten-year interval, after which the stand grows to the volume for the node at the end of the arc. These quantities are given in a table below Figure 9.1.

At age 30 the stand is clear felled and replanted and grows[3] again to state S_1. The objective is to choose the rotation that maximizes the amount of timber produced over the planning period of 30 years.

The optimal solution is as follows:

- Age 10: Take no action.

- Age 20: Execute a light thinning cut, removing 150 cubic meters per

---

[2]This quotation is from Dykstra [1984].

[3]Note that growth is not assumed to be linear, and this causes no difficulty in specifying the network. Linearity enters from the assumption that a decision action, for example, heavy thinning on, say, 14 hectares, will yield precisely twice the amount of timber obtained by heavy thinning on 7 hectares.

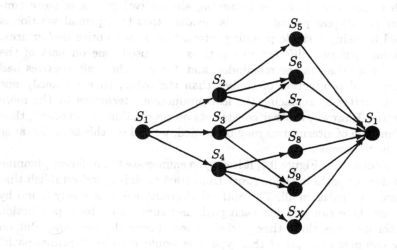

FIGURE 9.1. Network of alternatives for timber-harvesting example

State	Age	Volume
S_1	10	260
S_2	20	650
S_3	20	535
S_4	20	410
S_5	30	850
S_6	30	750
S_7	30	650
S_8	30	600
S_9	30	500
S_X	30	400

Transformation	Benefit
$S_1 \rightarrow S_2$	0.0
$S_1 \rightarrow S_3$	50.0
$S_1 \rightarrow S_4$	100.0
$S_2 \rightarrow S_5$	0.0
$S_2 \rightarrow S_6$	150.0
$S_2 \rightarrow S_7$	200.0
$S_3 \rightarrow S_6$	0.0
$S_3 \rightarrow S_7$	100.0
$S_3 \rightarrow S_9$	175.0
$S_4 \rightarrow S_8$	0.0
$S_4 \rightarrow S_9$	75.0
$S_4 \rightarrow S_X$	150.0
$S_5 \rightarrow S_1$	850.0
$S_6 \rightarrow S_1$	750.0
$S_7 \rightarrow S_1$	650.0
$S_8 \rightarrow S_1$	600.0
$S_9 \rightarrow S_1$	500.0
$S_X \rightarrow S_1$	400.0

hectare.

- Age 30: Clear fell, removing 750 cubic meters per hectare.

No other sequence of actions (planning alternative) produces more timber over the 30-year period. It is also evident that the optimal solution is achieved by using a single planning alternative on the entire timber area. For example, if two different alternatives were used, one on part of the area and the other on the remainder, and if one of these alternatives had a higher total production of timber than the other, then obviously one would do better by replacing the less productive alternative by the more productive. However, if other constraints on production are imposed, then a *combination* of alternatives may be needed; we will see this situation arise in the next example.

In the network of Figure 9.1, it is easy to enumerate the different planning alternatives or *paths* in the network from start to finish, and establish that there are nine paths in all. The optimal alternative is thus easily found by inspection. One can just list each path and sum up its total production of timber over the three time periods, then choose the best one. But on a more complex example of this type, one would need an "optimal-path" algorithm that is more efficient than enumeration. We will discuss solution strategies in Section 9.3.

9.2   A Rangeland Improvement Problem

Consider a 5000 acre tract of rangeland. It is presently covered with dense, mature chaparral,[4] but it has the potential of being turned into productive forage land that is suitable for the grazing of livestock, once the chaparral is eradicated and a grass cover is established. A strategic plan must be developed that schedules sequences of appropriate *decision actions* to effect this conversion over a 20-year planning period.

A representative acre of rangeland will be characterized by the following:

- chaparral cover, as a percentage of total cover;

- forage production, as a percentage of the potential forage production capacity;

- trend in chaparral cover.

---

[4]A type of weed, mentioned frequently in the news reports of the disastrous southern California wildfires of 2003.

For the purposes of the model, we will assume that each of the foregoing characteristics or *state parameters* can potentially assume a small number of different levels as follows:

1. chaparral cover: 100%, 30%, and 0%, which are identified by the values of the index $i = 0, 1, 2$, respectively;

2. forage production: 0%, 50%, and 100%, which are identified by the values of the index $j = 0, 1, 2$, respectively;

3. chaparral trend: no trend and positive trend,[5] which are identified by the values of the index $k = 0, 1$ respectively.

Many of these levels (i, j, k) cannot coexist, for example, $i = 0$, $j = 2$ (for obvious reasons) or $i = 0$, $k = 1$ (chaparral cover is already at a maximum). The set of valid combinations will define the set of potential *states* of the resource class. These are as follows:

$$\{S_1 = (0,0,0), \quad S_2 = (1,1,0), \quad S_3 = (1,1,1), \quad S_4 = (2,2,0)\}. \qquad (9.1)$$

Initially, the entire range resource is assumed to be in state $S_1 = (0,0,0)$.

A decision action is a prescription or practice during a 5-year planning interval that specifies how one acre of rangeland, which is in a given state from the above set (9.1) at the start of the interval, is transformed into another state at the end of the interval. It has an associated *cost* and *benefit* that are specified on a per-acre basis. The set of possible decision actions and their costs and benefits will be listed below.

A sequence of four (5-year) actions constitutes a strategic *decision alternative* or *activity* for the acre of rangeland under consideration, over the 20-year planning period.

A strategic *plan* for the rangeland resource prescribes an activity for each acre of the rangeland. Two different acres could obviously have the same prescribed activity, so equivalently we can say that a strategic plan partitions the 5000-acre tract and prescribes a particular activity for each partition. An *optimal plan* is sought that does not violate the *constraints* imposed on the rangeland resource decision problem and optimizes some quantified *objective* function.

The actions and their associated costs and benefits, the decision alternatives, and the constraints and objectives of the foregoing range resource decision problem will now be stated:

Actions: Each action is applicable to a subset of the set of states listed in (9.1). Given a "starting" state, a relevant action transforms it into a "destination" state, and this action has an associated cost and benefit,

---

[5]Positive trend means that the chaparral cover is on the increase. For simplicity, we ignore the possibility of negative trend.

TABLE 9.1. Transformations

Start	Action	Destin.	Cost	Ben.
(0,0,0)	No change (NC)	(0,0,0)	1.0	0.0
(0,0,0)	Spot burn, spray, seed (SBSPS)	(1,1,1)	8.0	5.0
(0,0,0)	Burn, spray, seed, respray (BSPSR)	(2,2,0)	12.0	10.0
(1,1,0)	Spot burn, spot seed (SBSS)	(2,2,0)	5.0	10.0
(1,1,1)	Spray (SP)	(1,1,0)	2.0	5.5
(2,2,0)	Maintain (M)	(2,2,0)	0.5	10.0

both measured in dollars/acre (benefit is the dollar value of forage that becomes available). This information is given in Table 9.1.

Alternatives: The set of decision alternatives for the foregoing range decision problem is defined by a network given in Figure 9.2. Each path in this network, leading from the initial state S_1 to any state at the end of the planning period, defines a valid decision alternative or activity. Note that not all states are attainable at the end of each interval; for example, the state S_2 is not attainable at the end of interval 1.

Constraints: The constraints on the resource decision problem are as follows: In each of the four 5-year planning intervals, 20 thousand dollars are available. In the four planning intervals, at least 10, 20, 30, and 40 thousand dollars respectively of benefit must be produced.

Objectives: Possible objectives are minimization of total cost over the planning period or maximization of total benefit over the planning period.

Let us choose the cost objective and seek to minimize it subject to the constraints stated. Then the optimal plans or decision alternatives along with the number of acres to which each is applied are shown in the upper part of Table 9.2 along with the final status of the resource in the lower table: states at the end of the planning horizon and the acreage in each state. The total (minimal) cost incurred is \$59,924, and the total benefit derived over the planning period is \$121,529. (Acreage and dollars are rounded to the nearest whole number in the optimal solution reported.)

Let us consider alternative scenarios in which environmental considerations come into play. The chaparral in state S_1 is known to be a fire hazard, and thus it is preferable that no acreage be left in this state at the end of the planning period. However, chaparral provides habitat for wildlife, and thus lower bounds on the acreage in states S_2 and S_3, where some chaparral is preserved, are imposed. Let us say that at the ends of intervals 2, 3, and 4, the number of acres in these states should be at least 500, 1000, and 1500, respectively. Again we minimize total cost. The optimal solution is shown in Table 9.3. Observe that additional actions come into play, the total (minimum) cost increases to \$64,683, and the total derived benefit is \$110,582. Suppose, on the other hand, that the total benefit is *maximized* subject to the environmental (and other previous) constraints. The optimal

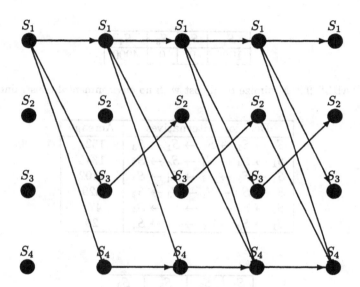

FIGURE 9.2. Network for the range improvement problem

State	Chaparral	Forage	Trend
S_1	100%	0%	none
S_2	30%	50%	none
S_3	30%	50%	positive
S_4	0%	100%	none

Transformation	Cost	Benefit
$S_1 \to S_1$	1.0	0.0
$S_1 \to S_3$	8.0	5.0
$S_1 \to S_4$	12.0	10.0
$S_2 \to S_4$	5.0	10.0
$S_3 \to S_2$	2.0	5.5
$S_4 \to S_4$	0.5	10.0

Decision Alternatives	Acreage
$S_1 \to S_1 \to S_4 \to S_4 \to S_4$	1426
$S_1 \to S_4 \to S_4 \to S_4 \to S_4$	1363
$S_1 \to S_1 \to S_1 \to S_4 \to S_4$	1211
$S_1 \to S_1 \to S_1 \to S_1 \to S_1$	1000

S_1	S_2	S_3	S_4
1000	0	0	4000

TABLE 9.2. Minimize total cost with no environmental constraints

Decision Alternatives	Acreage
$S_1 \to S_1 \to S_1 \to S_1 \to S_3$	1524
$S_1 \to S_4 \to S_4 \to S_4 \to S_4$	1345
$S_1 \to S_1 \to S_4 \to S_4 \to S_4$	1102
$S_1 \to S_1 \to S_1 \to S_3 \to S_2$	529
$S_1 \to S_1 \to S_3 \to S_2 \to S_4$	471
$S_1 \to S_3 \to S_2 \to S_4 \to S_4$	29

S_1	S_2	S_3	S_4
0	529	1524	2947

TABLE 9.3. Minimize total cost with environmental constraints

solution is shown in Table 9.4. The total incurred cost has increased further to $67,499, and the total (maximal) derived benefit is now $123,889. Many other scenarios can be envisioned and constraints imposed to achieve them, for example, limits on the fluctuations of expenditure and/or benefits over the period of planning.

9.3   Resource-Decision Software

The foregoing problems are two very simple examples of a general modeling-and-algorithmic approach that has broad applicability to the optimal development of natural and renewable resources in the presence of a variety of constraints on costs, benefits, environmental factors, and so on. This so-called $D_L P$ approach is described, in detail, in Nazareth [2001]. Computer software (on CD-ROM) is made available with this monograph for hands-on demonstration of the $D_L P$ decision support system, and it pro-

Decision Alternatives	Acreage
$S_1 \to S_1 \to S_1 \to S_3 \to S_2$	1417
$S_1 \to S_4 \to S_4 \to S_4 \to S_4$	1364
$S_1 \to S_1 \to S_4 \to S_4 \to S_4$	1107
$S_1 \to S_1 \to S_1 \to S_4 \to S_4$	529
$S_1 \to S_1 \to S_3 \to S_2 \to S_4$	500
$S_1 \to S_1 \to S_1 \to S_1 \to S_3$	83

S_1	S_2	S_3	S_4
0	1417	83	3500

TABLE 9.4. Maximize total benefit with environmental constraints

vides a very useful educational tool for experimentation with the foregoing timber and range models. The software makes it possible to specify a $D_L P$ model in an input language called DLPFI (pronounced "Delphi") that is easy to learn and use. For example, the input corresponding to the range-land planning model of Section 9.2 is shown in Table 9.5, and it is largely self-explanatory. Each line in the table that begins with a "?" is called a keyword record, and it corresponds to a main section of problem data. The latter is then specified on zero, one, or more subsequent lines of data, called data records, continuing up to the next keyword record. Thus, the problem is given a name, RANGE. It has 4 planning periods, or intervals. Only one resource class is specified, which is named CHAPR.[6] The states and actions for class CHAPR are identified by a unique name (of up to 8 characters). The state (i, j, k) is given the name STijk, and the list of all STATES is specified. This is followed by a list of all ACTIONS used (see the second column of Table 9.1). The next section of the input specifies the NETWORK of planning alternatives, and it reflects the information given in Table 9.1. The LOCAL constraints section specifies the INITIAL condition of the class as consisting of 5000 acres. (Other more general constraints on desirable and undesirable states in each interval can also be provided here.) The GLOBAL constraints section specifies the upper limits on cost, the lower limits on benefit, and identifies two possible objectives.[7] The OB-JECTIVE section selects the cost minimization objective. For full detail, see Nazareth [2001].

The corresponding output is shown in Table 9.6. The ALTERNATIVES

---

[6]This name is short for "chaparral." A more general problem could involve simultaneous planning for several resource classes.

[7]C[:] is equivalent to C[CHAPR:1] + C[CHAPR:2] + $\cdots$ + C[CHAPR:4], i.e., omitted names or periods implicitly range over all possibilities.

```
?PROBLEM    RANGE
?PERIODS
   4
?CLASSES
  CHAPR
?NAME  CHAPR
?STATES   CHAPR
  ST000  ST110  ST111    ST220
?ACTIONS CHAPR
 NC SBSPS BSPSR SBSS SP M
?NETWORK CHAPR
  ST000 NC 1.0 0.0 ST000
  ST000 SBSPS 8.0 5.0 ST111
  ST000 BSPSR 12.0 10.0 ST220
  ST110 SBSS 5.0 10.0 ST220
  ST111 SP 2.0 5.5 ST110
  ST220 M 0.5 10.0 ST220
?LOCAL CHAPR
?INITIAL CHAPR
  ST000=5000.
?CONSTRAINTS CHAPR
?GLOBAL
  C[CHAPR:1]<=20000.
  C[CHAPR:2]<=20000.
  C[CHAPR:3]<=20000.
  C[CHAPR:4]<=20000.
  B[CHAPR:1]>=10000.
  B[CHAPR:2]>=20000.
  B[CHAPR:3]>=30000.
  B[CHAPR:4]>=40000.
  C[:]=ZCOST
  B[:]=ZBEN
?OBJECTIVE
  MINIMIZE ZCOST
?ENDATA
```

TABLE 9.5. Sample DLPFI input for rangeland model

section gives the quantity that is managed by each alternative in the optimal solution, in terms of both proportion and total number of acres (the last number in the line containing the INITIAL keyword); the corresponding sequence of actions and their associated costs and benefits per acre; and the state attained after each action. The GLOBAL CONSTRAINTS section gives the values of the right-hand sides and the associated values of the slack variables (the amount by which a constraint fails to be satisfied as an equality, which is a nonnegative number for a $\leq$ constraint and a nonpositive number for a $\geq$ constraint). The OBJECTIVE section gives the optimal objective value of 59923.6 dollars. One can see that these optimal choices match the tabulation given in Section 9.2.

The DLPFI input language makes it possible to run a wide variety of scenarios with great ease, for example, the ones described at the end of Section 9.2. Each requires very simple modifications to the input in Table 9.5. The environmental constraints setting lower limits on acreage in states with an intermediate amount of chaparral can be specified by adding the following data records in the ?LOCAL section after the ?CONSTRAINTS keyword record:

```
?CONSTRAINTS CHAPR
   4: ST000 <= 0.
   2: ST110 + ST111 >= 500.
   3: ST110 + ST111 >= 1000.
   4: ST110 + ST111 >= 1500.
```

The number at the beginning of each data record identifies a period for which the constraint applies. The remainder of the data record then specifies the states involved in the constraint and the right-hand side level. Likewise, the change of objective when total benefit is maximized is specified after the ?OBJECTIVE keyword as follows:

```
?OBJECTIVE
   MAXIMIZE ZBEN
```

where the variable that is maximized is defined in the prior ?GLOBAL section. For further detail on the DLPFI language, see Nazareth [2001].

$D_L P$ models are an important subclass of the general linear programming approach and are useful in settings where time-staged decision making and an underlying network structure (associated with the discrete states of a dynamical system) play a prominent role. In traditional dynamic-linear programming, the modeling and algorithmic focus is on the equivalent LP formulation and explicit generation of the alternatives. (This LP loses the pure network structure described in Chapter 4 in the presence of side constraints, but it continues to be large and sparse.) In the $D_L P$ approach, the focus is on retaining the network formulation throughout and integrat-

```
?PROBLEM    RANGE
?NAME  CHAPR
?ALTERNATIVES         4
?CLUSTER    2.72727E-01
?INITIAL  ST000         5.00000E+03      1.36364E+03
  BSPSR          1.20000E+01      1.00000E+01      ST220
  M              5.00000E-01      1.00000E+01      ST220
  M              5.00000E-01      1.00000E+01      ST220
  M              5.00000E-01      1.00000E+01      ST220
?CLUSTER    2.42149E-01
?INITIAL  ST000         5.00000E+03      1.21074E+03
  NC             1.00000E+00      0.00000E+00      ST000
  NC             1.00000E+00      0.00000E+00      ST000
  BSPSR          1.20000E+01      1.00000E+01      ST220
  M              5.00000E-01      1.00000E+01      ST220
?CLUSTER    2.85124E-01
?INITIAL  ST000         5.00000E+03      1.42562E+03
  NC             1.00000E+00      0.00000E+00      ST000
  BSPSR          1.20000E+01      1.00000E+01      ST220
  M              5.00000E-01      1.00000E+01      ST220
  M              5.00000E-01      1.00000E+01      ST220
?CLUSTER    2.00000E-01
?INITIAL  ST000         5.00000E+03      1.00000E+03
  NC             1.00000E+00      0.00000E+00      ST000
  NC             1.00000E+00      0.00000E+00      ST000
  NC             1.00000E+00      0.00000E+00      ST000
  NC             1.00000E+00      0.00000E+00      ST000
?LOCAL CONSTRAINTS         0
?GLOBAL CONSTRAINTS        10
    0.00000E+00    L           2.00000E+04
    0.00000E+00    L           2.00000E+04
    3.07645E+03    L           2.00000E+04
    1.70000E+04    L           2.00000E+04
   -3.63636E+03    G           1.00000E+04
   -7.89256E+03    G           2.00000E+04
   -1.00000E+04    G           3.00000E+04
    0.00000E+00    G           4.00000E+04
   -5.99236E+04    ZCOST       0.00000E+00
   -1.21529E+05    ZBEN        0.00000E+00
?OBJECTIVE         9
    5.99236E+04
?ENDATA
```

TABLE 9.6. Sample output for rangeland model

ing techniques of dynamic programming[8] and linear programming at both the modeling and algorithmic levels. The resulting "whole" is greater than "the sum of the component parts," hence the use of a new acronym $D_L P$ to distinguish it.

Exercise: Formulate the timber and rangeland problems explicitly as linear programs in two different ways:
a) by enumerating the complete set of alternatives (paths in the network) with associated variables corresponding to the amounts of the resource (numbers of hectares or acres of land) used with each alternative, and then writing out the associated objective and constraints.
b) by defining variables to be the flows (number of hectares or acres transformed) between nodes, and writing out the flow balance and other constraints.

9.4   Notes

Sec 9.1: The timber example and quotation are from Dykstra [1984], as is the earlier "close-to-home" version described in Section 1.3.

Sec 9.2: This example is from the pioneering rangeland model of Jansen [1974].

Sec 9.3: For much more detail and a wide variety of applications, see Nazareth [2001]. The hands-on $D_L P$ demonstration diskette that is also provided with the monograph is easy to install on a PC and can be used to experiment with the $D_L P$ models described in Sections 9.1 and 9.2. This educational tool is derived from a prototype implementation of the $D_L P$ system written in Fortran. *In order to achieve maximal portability*, it is compiled for MS-DOS—the most basic PC system—and it uses a very simple user interface. This permits the program to be run, as well, on any PC system under Windows, for example, Windows 3.1, NT, Me, XT, and so on. The recommended approach is explicitly to open an MS-DOS window. On older Windows systems, an icon is provided for this purpose on the desktop. On more recent systems, proceed from "Start" via the "Accessories" and "Command Prompt" items (Windows XT), or the "Programs" followed by "Applications" and "MS-DOS" menu items (Windows Me). Once the command window is open, continue as described in the appendix of Nazareth [2001]. Alternatively, "Run" the program from the "Start" menu, or use

---

[8]This is an important optimization technique developed by R. Bellman that is not discussed in this primer. For details, see Bellman and Dreyfus [1962] or other more recent optimization texts. In particular, the example of Section 9.1, a longest-path problem, can be solved by a pure dynamic programming recursion as described by Dykstra [1984].

the File manager (also called Windows Explorer) and click the DLPDEMO
file name, after it has been copied from the CD-ROM to a new directory,
say on the hard drive. A user familiar with Windows will have no difficulty
finding the most convenient among the foregoing set of options for running
the DLPDEMO program as an educational tool.

10
Optimization: The Big Picture

The visual schematics given earlier in Figures 4.2 and 8.1 provide a "taxonomy" of the optimization models and algorithms considered in this primer.[1] Their purpose is to give the reader an overall sense of the optimization landscape and the interrelationships between its prominent features. But of course, they should not be carved in stone.

In this final chapter, we provide a third visual schematic that summarizes the "big picture" of optimization and brings our primer to a conclusion.

10.1 A "Cubist" Portrait

The rangeland improvement problem of the preceeding chapter, Section 9.2, provides an excellent vehicle for describing some of the key partitions within the field of optimization, in particular, the following:

- *Continuous vis-à-vis Discrete*: The variables in the rangeland improvement example are the numbers of acres that are managed by different planning alternatives. These variables do not have to be whole numbers, and they can take *any* value between 0 acres and the total of 5000 acres; i.e., they are continuous, real-valued variables. Suppose we insist that the variables must be integers or that the entire acreage must be managed by a single alternative. Then

---

[1]Taxonomy is defined in Webster's Dictionary as an "orderly classification of plants or animals according to their presumed natural relationships."

we have entered the realm of discrete optimization (integer; (0,1); combinatorial). Such models are often much more difficult to solve than continuous models. Interestingly enough, the subject of network flows occupies the *middle ground* between continuous and discrete optimization. The network variables are permitted to assume continuous values (real numbers). But at optimality they are automatically integer valued, without the need to incorporate such restrictions *explicitly*, whenever the exogenous data values of the underlying (pure) network model are integral. We saw examples of this property in Chapter 2 in the discussion of the maximum flow model, and it is true of many other network flow models.

- *Deterministic vis-à-vis Stochastic*: Suppose the outcome of actions in the rangeland example are uncertain. For example, consider an acre in state $S_3 = (1,1,1)$ under action SP. Whereas previously we assumed that the resulting state was $S_2 = (1,1,0)$ with certainty, let us now assume that with probability 0.8 the action is successful in reducing the third parameter from "positive trend" to "no trend," and with probability 0.2 the action is not successful and the ensuing state remains S_3. Similar probabilistic outcomes can occur for some of the other actions. The resulting model must then seek to optimize *expected* values of costs and benefits, and constraints must be formulated likewise. The optimization model has entered the probabilistic realm, also known as *stochastic programming*.

- *Finite-Dimensional (F-D) vis-à-vis Infinite-Dimensional (∞-D)*: Suppose the state *parameters* in the rangeland improvement model are allowed to take any admissable values. For example, chaparral cover can be any number between 0 and 100%. Then, instead of a small finite number of states, we have an infinite number of possible states. Suppose also that instead of making planning decisions at the start of just four (5-year) planning intervals, we can employ continuous control actions over the time dimension, or planning period, of 20 years. Then we have entered the realm of continuous *optimal control*, i.e., optimization over a space of control functions, or infinite-dimensional optimization. This again opens fresh optimization challenges that build extensively on finite-dimensional optimization techniques.

If these three key partitions are represented on three axes and the special role of networks in mediating between the continuous and discrete cases is also highlighted, then we obtain the schematic cube shown in Figure 10.1. In much of the discussion in this booklet, the focus has been on optimization models and algorithms *within the deterministic, continuous, finite-dimensional "subcube" and the network-flows "slice" immediately below it*. As we noted in Chapter 8, this "subcube" covers both smooth, or differentiable, and nonsmooth, or nondifferentiable, optimization, and the visual schematic of Figure 8.1 could indeed be used for problems in both areas, although the focus there was on problems defined in terms of differentiable objectives and constraints. Similarly, the visual schematic of Figure 4.2 addresses problems within a portion of the network "slice," as mentioned above. These two schematics summarize substantial problem areas that involve many different types of optimization models and algorithms with wide applicability. *But it is evident from Figure 10.1 that optimization encompasses a very much larger area.* Some of the problems have been touched on in the foregoing itemized discussion, in particular, stochastic programming, integer programming, and optimal control. For the interested reader, an overview of the entire field can be found in the recently published *Encyclopedia of Optimization* edited by Floudas and Pardalos [2001].

As we have noted in earlier chapters, the exploration of optimization began in earnest almost six decades ago with the modeling-and-algorithmic revolution fathered by G.B. Dantzig—see, in particular, the selection of his key contributions given in Cottle [2003]—and it is thus appropriate to close with a quotation from Dantzig's foreword in the recently published textbook of Dantzig and Thapa [1997, 2003]:

> Linear programming[2] can be viewed as part of a great revolutionary development that has given mankind the ability to state general goals and to lay out a path of detailed decisions to be taken in order to "best" achieve these goals when faced with practical situations of great complexity. Our tools for doing this are ways to formulate real-world problems in detailed mathematical terms (models), techniques for solving the models (algorithms), and engines for executing the steps of algorithms (computers and software).

And in concluding his assessment of the state of the art of the field:

> *The ability to state general objectives and then be able to find optimal policy solutions to practical decision problems of great complexity is the revolutionary development I spoke of earlier.* We have come a long way down the road to achieving this goal, but much work remains to be done, particularly in the area of uncertainty. The final test will come when we can solve the practical problems under uncertainty that originated the field back in 1947.

---

[2]And, indeed, *all* of mathematical programming.

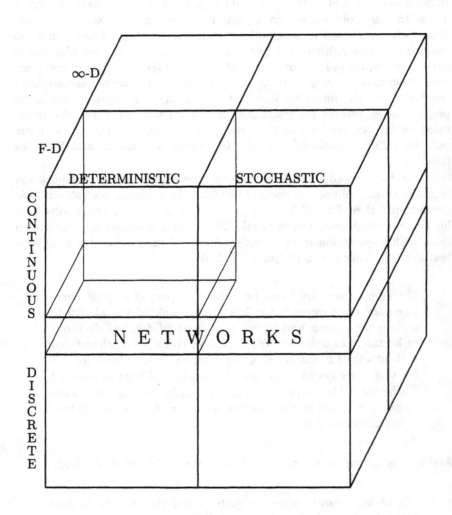

FIGURE 10.1. A "Cubist" portrait of optimization

10.2   Notes

Section 10.1: The bibliography lists several classic references, as well as more recent articles and books on optimization; see, in particular, the landmark series of monographs from Princeton University Press: in chronological order, Von Neumann and Morgenstern [1944], Ford and Fulkerson [1962], Dantzig [1963], and Rockafellar [1970]. For other key references, see the extensive bibliographies of the works cited. And for a historical perspective, the essays in the book edited by Lenstra et al. [1991] provide some delightful accounts, written by pioneers in the field of optimization.

Only a minimal level of mathematical background has been assumed in this primer, and the *mathematics of optimization* has been left largely unaddressed. Some *key topics and classic references* on this foundational activity in optimization, organized under the three main themes of our primer's subtitle, are as follows:

Models: *Convex analysis* of optimization models, namely, the study of smooth and nonsmooth convex functions and sets, and of optimization models that are based on them, is fundamental to optimization, and the subject was brought to flower in the classic work of Rockafellar [1970].

Algorithms: Optimization and equation-solving algorithms remain *heuristic* until their convergence and numerical stability have been formally established. Four key aspects of the subject are:

- *global convergence* analysis of an algorithm from an arbitrary starting point to a solution of the optimization or equation-solving problem; see Ortega and Rheinboldt [1970];
- *local rate of convergence* analysis of an algorithm from a point in the near vicinity of a solution to the solution itself; see again Ortega and Rheinboldt [1970];
- *global rate of convergence*, or *complexity*, analysis of an algorithm, from an arbitrary starting point to a solution; see Blum et al. [1998];
- *numerical stability* analysis of an algorithm as a prelude to effective implementation and the development of high-quality software; see Wilkinson [1963].

Duality: Ford–Fulkerson duality, introduced in Chapter 2, and König–Egerváry duality, introduced in Chapter 3, are the fountainhead for a broad stream of network flow duality results encompassing models and algorithms of max-flows/min-cuts, matching/covering, transportation, and assignment; see Figure 4.2. A second stream of duality results for linear programming, whose source is Farkas [1902], was introduced in Chapter 5. These two streams of duality results share a broad common frontier, because network flow problems can be expressed explicitly as linear programs, which, in turn, can be dualized. For example, the linear program equivalent to the maximum flow example of Chapter 2, which was formulated in Section 4.1, has an associated dual linear program. It follows from the relationships in Figure 4.2 that max-flow/min-cut (Ford–Fulkerson) and max-matching/min-covering (König–Egerváry) duality, which were developed within

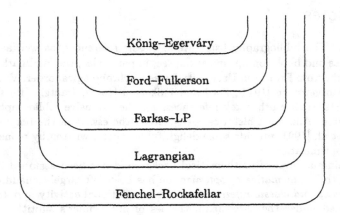

FIGURE 10.2. Nested duality results

the context of networks in Sections 2.3 and 3.3, can alternatively be inferred from Farkas–LP duality applied to equivalent linear programs. In other words, Farkas–LP duality subsumes Ford–Fulkerson duality, and the latter, in turn, subsumes König–Egerváry duality.

Farkas–LP duality is itself subsumed by so-called *Lagrangian duality* for non-linear programming, which has roots in classical Lagrangian theory for nonlinear equality constrained problems. For a concise and readable development, see Zangwill [1969]. And these duality reults are further subsumed by so-called *conjugate duality* theory, which is based on the pioneering works of Fenchel [1949] and Rockafellar [1974], and has a much broader expression within the setting of convex and general nonlinear programming, both smooth and nonsmooth.

The foregoing observations concerning nested manifestations of the duality principle are summarized by Figure 10.2.

References

[1] Avriel, M. (1976), *Nonlinear Programming: Analysis and Methods*, Prentice-Hall, Englewood Cliffs, New Jersey.

[2] Bachem, A., Grötschel, M. and Korte, B. (eds.) (1983), *Mathematical Programming: The State of the Art, Bonn 1982*, Springer-Verlag, Berlin.

[3] Bazaraa, M.S., Jarvis, J.J., and Sherali, H.D. (1990), *Linear Programming and Network Flows*, Second Edition, Wiley, New York.

[4] Bellman, R. and Dreyfus, S. (1962), *Applied Dynamic Programming*, Princeton University Press, Princeton, New Jersey.

[5] Bertsekas, D.P. (1999), *Nonlinear Programming*, Athena Scientific, Belmont, Massachusetts (Second Edition).

[6] Bland, R.G. (1977), "New finite pivoting rules for the simplex method," *Mathematics of Operations Research*, 2, 103–107.

[7] Blum, L., Cucker, F., Shub, M., and Smale, S. (1998), *Complexity and Real Computation*, Springer-Verlag, New York (with a foreword by R.M. Karp).

[8] Brent, R.P. (1973), *Algorithms for Minimization without Derivatives*, Prentice Hall, Englewood Cliffs, New Jersey (republished by Dover Publications in 2002).

[9] Chvátal, V. (1983), *Linear Programming*, W.H. Freeman and Company, New York.

[10] Cottle, R.W. (ed.) (2003), *The Basic George B. Dantzig*, Stanford University Press, Palo Alto, California.

[11] Dantzig, G.B. (1963), *Linear Programming and Extensions*, Princeton University Press, Princeton, New Jersey.

[12] Dantzig, G.B. (1980), "Expected number of steps of the simplex method for a linear program with a convexity constraint," Technical Report SOL 80-3, Systems Optimization Laboratory, Department of Operations Research, Stanford University, Stanford, California.

[13] Dantzig, G.B. (1983), "Reminiscences about the origins of linear programming," in *Mathematical Programming: The State of the Art, Bonn, 1982*, A. Bachem, M. Grötschel, and B. Korte (eds.), Springer-Verlag, Berlin, 78-86.

[14] Dantzig, G.B. (1985), "Impact of linear programming on computer development," Technical Report SOL 85-7, Systems Optimization Laboratory, Department of Operations Research, Stanford University, Stanford, California.

[15] Dantzig, G.B. and Thapa, M.N. (1997, 2003), *Linear Programming. I: Introduction, II: Theory and Extensions*, Springer Series in Operations Research, Springer-Verlag, New York.

[16] Dikin, I.I. (1967), "Iterative solution of problems of linear and quadratic programming," *Soviet Mathematics Doklady*, 8, 674-675.

[17] Dykstra, D.P. (1984), *Mathematical Programming for Natural Resource Management*, McGraw-Hill, New York.

[18] Edmonds, J. and Karp, R.M. (1972), "Theoretical improvements in algorithmic efficiency for network flow problems," *Journal of the ACM*, 19, 248-264.

[19] Egerváry, J. (1931), "Matrixok kombinatorikus tulajdonságairól," *Mathematikai ès Fizikai Lápok*, 38, 16-28. Translation by H.W. Kuhn (1955), "On combinatorial properties of matrices," George Washington University Logistics Paper 11.

[20] Evans, J.R. and Minieka, E. (1992), *Optimization Algorithms for Networks and Graphs*, Second Edition Revised and Expanded, Marcel Dekker, New York.

[21] Farkas, J. (1902), "Über die theorie der einfachen ungleichungen," *Journal für die Reine und Angewandte Mathematik*, 124, 1-24.

[22] Fenchel, W. (1949), "On conjugate convex functions," *Canadian J. Mathematics*, 1, 73-77.

[23] Fiacco, A.V. and McCormick, G.P. (1968), *Nonlinear Programming: Sequential Unconstrained Minimization Techniques*, Wiley, New York. (Republished in Classics in Applied Mathematics Series, Volume 4, SIAM, Philadelphia, 1990.)

[24] Floudas, C.A. and Pardalos, P.M., (eds.) (2001), *Encyclopedia of Optimization*, Vols. I–VI, Kluwer Academic Publishers, Dordrecht and Boston.

[25] Ford, L.R., Jr., and Fulkerson, D.R. (1962), *Flows in Networks*, Princeton University Press, Princeton, New Jersey.

[26] Frisch, K.R. (1955), "The logarithmic potential method for convex programming," manuscript, Institute of Economics, University of Oslo, Oslo, Norway.

[27] Frisch, K.R. (1956), "La résolution des problèmes de programme lineaire par la méthode du potential logarithmique," *Cahiers du Séminaire D'Econométrie*, 4, 7–20.

[28] Garey, M.R. and Johnson, D.S. (1979), *Computers and Intractability: A Guide to the Theory of NP-Completeness*, W.H. Freeman and Company, San Francisco and New York.

[29] Huard, P. (1967), "Resolution of mathematical programming with nonlinear constraints by the method of centers," in *Nonlinear Programming*, J. Abadie (ed.), North Holland, Amesterdam, 207–219.

[30] Jansen, H.C. (1974), *A Resources Allocation Method for Range Resource Management*, Ph.D. Dissertation, University of California, Berkeley, California.

[31] Kantorovich, L.V. (1939), "Mathematical methods in the organization and planning of production." [English translation: *Management Science*, 6, 366–422 (1960).]

[32] Karmarkar, N. (1984), "A new polynomial-time algorithm for linear programming," *Combinatorica*, 4, 373–395.

[33] Khachiyan, L.G. (1979), "A polynomial algorithm in linear programming," *Soviet Mathematics Doklady*, 20, 191–194.

[34] Komornik, V. (1998). "A simple proof of Farkas' lemma," *American Mathematical Monthly*, 105, No. 10, 949–950.

[35] König, D. (1936), *Theorie der Endlichen und Unendlichen Graphen*, Akademische Verlagsgesellschaft, Leipzig. Reprint 1950, Chelsea.

[36] Koopmans, T.C. (ed.) (1951), *Activity Analysis of Production and Allocation*, Cowles Commission Monograph 13, Wiley, New York.

[37] Kozen, D.C. (1992), *The Design and Analysis of Algorithms*, Springer-Verlag, New York.

[38] Lawler, E. (1976), *Combinatorial Optimization: Networks and Matroids*, Dover Publications, Mineola, New York (reprint, 2001).

[39] Lenstra, J.K., Rinnooy Kan, A.H.G., and Schrijver, A. (eds.) (1991), *History of Mathematical Programming: A Collection of Personal Reminiscences*, North-Holland, Amsterdam, The Netherlands.

[40] Luenberger, D.G. (1984), *Linear and Nonlinear Programming*, Addison-Wesley (Second Edition).

[41] McLinden, J. (1980), "The analogue of Moreau's proximation theorem, with applications to the nonlinear complementary problem," *Pacific Journal of Mathematics*, 88, 101–161.

[42] Megiddo, N. (1989), "Pathways to the optimal set in linear programming," in *Progress in Mathematical Programming: Interior-Point and Related Methods*, N. Megiddo (ed.), Springer-Verlag, New York, 131–158.

[43] Nazareth, J.L. (1987), *Computer Solution of Linear Programs*, Oxford University Press, Oxford and New York.

[44] Nazareth, J.L. (1994), *The Newton–Cauchy Framework: A Unified Approach to Unconstrained Nonlinear Minimization*, Lecture Notes in Computer Science, Vol. 769, Springer-Verlag, Heidelberg and Berlin.

[45] Nazareth, J.L. (2001), $D_L P$ *and Extensions: An Optimization Model and Decision Support System*, Springer-Verlag, Heidelberg and Berlin.

[46] Nazareth, J.L. (2003), *Differentiable Optimization and Equation Solving: A Treatise on Algorithmic Science and the Karmarkar Revolution*, Springer-Verlag, New York.

[47] Nesterov, Y.E. and Nemirovskii, A.S. (1994), *Interior Point Polynomial Algorithms in Convex Programming*, SIAM Studies in Applied Mathematics, Vol. 13, SIAM, Philadelphia.

[48] Ortega, J.M. and Rheinboldt, W.C. (1970), *Iterative Solution of Nonlinear Equations in Several Variables*, Academic Press, New York.

[49] Renegar, J. (2001), *A Mathematical View of Interior-Point Methods in Convex Programming*, SIAM, Philadelphia.

[50] Rockafellar, R.T. (1970), *Convex Analysis*, Princeton University Press, Princeton, New Jersey.

[51] Rockafellar, R.T. (1974), *Conjugate Duality and Optimization*, Regional Conference Series in Applied Mathematics 16, SIAM, Philadelphia.

[52] Rockafellar, R.T. (1998), *Network Flows and Monotropic Optimization*, Athena Scientific, Belmont, Massachusetts (originally published by J.H. Wiley and Sons, New York, in 1984).

[53] Sonnevend, G. (1986), "An 'analytic center' for polyhedrons and new classes of global algorithms for linear (smooth, convex) programming," in A. Prekopa, J. Szelezsan, and B. Strazicky (eds.). *System Modelling and Optimization*, Lecture Notes in Control and Information Sciences, Vol 84, Springer-Verlag, Heidelberg and Berlin, 866–875.

[54] Traub, J.F. (1964), *Iterative Methods for the Solution of Equations*, Prentice-Hall, Englewood Cliffs, New Jersey.

[55] Von Neumann, J. and Morgenstern, O. (1944), *Theory of Games and Economic Behaviour*, Princeton University Press, Princeton, New Jersey.

[56] Wilkinson, J.H. (1963), *Rounding Errors in Algebraic Processes*, Prentice-Hall, Englewood Cliffs, New Jersey.

[57] Winston, W.L. (1997), *Operations Research*, Wadsworth Publishing Company, Belmont, California (Third Edition).

[58] Zangwill, W.I. (1969), *Nonlinear Programming: A Unified Approach*, Prentice Hall, Englewood Cliffs, New Jersey.

Index

ABOUT THE AUTHOR

For a more complete biographical sketch and list of publications, see:
http://www.math.wsu.edu/faculty/nazareth

John Lawrence (Larry) Nazareth was educated at the University of Cambridge (Trinity College), England, and the University of California at Berkeley. He is author of over seventy-five technical articles, and he has written five other books on optimization as follows: *Computer Solution of Linear Programs* (Oxford University Press, 1987); *The Newton–Cauchy Framework: A Unified Approach to Unconstrained Nonlinear Minimization* (Springer-Verlag, 1994); *Linear and Nonlinear Conjugate-Gradient Related Methods* (a coedited volume, SIAM, Philadelphia, 1996); $D_L P$ *and Extensions: An Optimization Model and Decision Support System* (Springer-Verlag, 2001; includes a CD-ROM containing an executable version of a prototype implementation comprising several thousands lines of Fortran code); *Differentiable Optimization and Equation Solving: A Treatise on Algorithmic Science and the Karmarkar Revolution* (Springer-Verlag, 2003). He has also authored two short nontechnical books for limited circulation: *Three Faces of God and Other Poems* (Apollo Printing, Berkeley, 1986), and *Reminiscences of an Ex-Brahmin: Portraits of a Journey Through India* (Apollo Printing, Berkeley, 1996); and he is, at present, completing a collection of essays: *Thou Art That and Other Philosophical Reflections*.